D1235865

GROWING AND COOKING BEANS

GROWING AND COOKING

BEANS

by John E. Withee

YANKEE, Inc.
Dublin, New Hampshire

This book has been prepared by the Staff of Yankee, Inc.
Designed by Carl Kirkpatrick
Illustrated by Ray Maher

Published 1980
Yankee, Inc., Dublin, New Hampshire 03444

First Edition
Second Printing, 1982
Copyright 1980, by Yankee, Inc.
Printed in the United States of America

Library of Congress Catalog Card No. 79-57179
ISBN 0-911658-05-X

DEDICATION

I dedicate this book to my
wife, Ruth, who has put up
with me for all these years.

ACKNOWLEDGMENTS

I would like to thank the following people for their help and support.

All the subscribers to my newsletter, *The Wanigan,* especially those who sent in the recipes that appear in this book, and those who donated their heirloom beans.

Albert Erlon Mosher of Gorham, Maine, and the oldest farm in Maine still working under the same family, who was responsible for my early interest in beans.

Professor Roger F. Sandsted in the Department of Vegetable Crops at Cornell University, for encouraging me to start a collection of heirloom beans.

Professor Elwyn Meader of Meaderboro, New Hampshire, for his expertise and advice in many areas.

And special thanks to my daughter, Lianne Carlin, for her hours of typing help.

GROWING AND COOKING BEANS

CONTENTS

GROWING AND COOKING BEANS

INTRODUCTION

INTRODUCTION

The world has been full of beans for thousands of years. They have been a mainstay of the food supply for probably as long as man has cared for plants and practiced agriculture. This is not surprising, since beans have many virtues not found in other food plants. In the garden and on the table, the bean is noted for its adaptability and its common appeal. From an easily grown crop of delicate "French" beans to the culinary triumph of a cassoulet, we have in the bean the one vegetable that brings a "rich man's food," to quote a sixteenth-century writer, to the tables of every gastronome, poor or rich, and in variety.

Botanically classed in the Family of plants named *Leguminosae,* which means pod bearing, beans are similar to peas and other legumes in their ability, given the right conditions, to capture nitrogen from the air and use it to aid their growth. Even the earliest agricultural accounts recognized the value of these plants for enriching the soil.

The bean Family seems to have had its wild beginnings in the world's tropical regions. Early humans extended the range of bean growth to sub-tropic and temperate areas where our more modern men adopted them. As these food beans became more important, learned people commenced to identify, separate, and "class" them. Changes seem to have been the rule in botany, but it is interesting to note how things stand today regarding the more detailed classification in beans, which is the genus (pl. genera). Here is a list of the more common beans of the world with their native homes and botanical names. (Question marks indicate supposed, but not certain, origins.)

Table 1.

Common Name	Botanical Name	Native Home
Adsuki bean (Adzuki)	*Vigna angularis*	Asia, Japan
Asparagus bean (Yard-long bean)	*Vigna sesquipedalis*	? Asia
Blackeye bean (Black-eyed pea)	*Vigna sinensis*	East Indies
Bonavist bean (Hyacinth bean)	*Dolichos lablab*	Old World Tropics
Broad bean (Horsebean, Fava)	*Vicia faba*	? Africa
Chickasaw lima (Jack bean, Horsebean)	*Canavalia ensiformis*	Tropical Africa
Kidney bean (Common bean)	*Phaseolus vulgaris*	Tropical Americas
Lima bean	*Phaseolus lunatus*	Tropical Americas
Mung bean	*Vigna radiata*	? India
Rice bean	*Vigna umbellata*	Asia
Runner bean	*Phaseolus coccineus*	Tropical Americas
Soybean	*Glycine max*	Manchuria — Japan
Sword bean	*Canavalia gladiata*	Old World Tropics
Tepary bean	*Phaseolus acutifolius*	Mexico
Winged bean	*Psophocarpus tetragonolobus*	Tropics

At least some of the wild ancestors of these beans are still living in remote areas and are valuable for plant breeding purposes. The genus *Phaseolus,* by far the most widely cultivated of the edible beans, is the focus of this book. These native American beans consist of many diverse forms, some very old.

Botanical archaeologists, since the advent of the Carbon-14 radioisotope dating technique, have been able to determine that certain *Phaseolus* beans were stored in caves in Peru in 5000 B.C. Other types of beans found in Mexican caves have been dated back three thousand years, and caches of bean seed from cliff dweller sites in our Southwest date back to A.D. 1000. These American beans, and apparently none from other continents, had spread by Indian trading to much of the present United States, the Caribbean Islands, and eastern South America long before the "discoveries" of these lands by Europeans.

It is not known exactly which of the numerous types were being grown by the various Indian groups before the white incursion. What does seem likely is that the preferences of today may be continuations of older preferences. So, the Caribbean people prefer Black Turtle Soup beans, as do the people of Brazil. The Southwest favors the Pinto and Mexican Red. Red beans are old favorites in the South. Boston seems to think Pea beans are just right. Undoubtedly, the variety in beans was here, ready for Europeans to take home, which Columbus did. It must have been a pleasant surprise to the explorer to find a bean superior to his Broad beans!

By 1601, *Phaseolus* beans were spread over Europe and had become, according to Parkinson in 1629, "a common, delicate food." It is not clear what, if any, type of bean exchange took place in the Americas during the Spanish incursions, which preceded the well-documented English period.

Curiously enough, our 1607 and 1620 English settlers seemed to find the Indian methods of planting beans to be worth noting, with no mention being made of bringing their own European *Phaseolus* seeds. In any case, the beans had now come full circle. Not only were the native American beans an improvement in quality, they were also well adapted to the climate and soils, after centuries.

In these modern times American beans are classed in the *Phaseolus* genera, as mentioned, and are further separated by species:

Phaseolus coccineus is commonly known as the Scarlet Runner bean. It also has a subspecies, alba, the White Runner bean. Scarlet Runner, a large seed with black streaks over purple, is generally grown for its bright red blossoms. It is a climbing plant as is the White Runner, which bears white bloom and white seed. Both are edible.

Phaseolus lunatus is the lima bean, so named for its Lima, Peru, South America heritage. Limas are classed as Large Lima, having rather flat white or colored seed, Potato Lima, with thick seed (somewhat like the White Runner), and the Sieva or Small Lima, which is found in colors also. All have both climbing and bush strains.

Phaseolus vulgaris is the "kidney" bean. It is the species having the greatest number of varieties or subspecies, greatest size variations, most color patterns, and most uses in its many edible forms, pods to dry seed. Growth patterns of this species range from dwarf through runner forms, to full, vigorous climbing sorts.

These three *Phaseolus* species are known to have over four thousand strains. At least some of each species are adapted to growing in most areas of the United States, and many are excellent food when consumed in any of their growth stages.

Because of this diversity in form, texture, color, and taste, every growing and eating requirement or desire among bean lovers can be met by concentrating on American *Phaseolus*.

John E. Withee
Lynnfield, Massachusetts
January 1980

GROWING AND COOKING BEANS

GROWING BEANS

GROWING BEANS

The bean plant is not generally considered to be difficult or demanding for the general gardener. Compared to celery, or perhaps mushrooms, bush snap beans are definitely easier for the beginning grower to produce. However, compared with radishes or weeds, beans do need a bit more care and can be a bit of a chore. Sometimes we need to be armed with at least a forewarning of troubles that are lurking and the knowledge to forestall them.

Insects and diseases are only two of the obstacles which bean producers face every season. The necessary research for lessening these hindrances requires workers whose training and life work is zeroed in on specific problems. They are specialists, and the results of their studies are revealed in over two hundred reports, papers, etc., which are published every month, internationally, just on the subject of beans.

The complexity of it is perhaps best shown in this list of some of the titles of these bean specialists:

Plant Pathologist
Plant Breeder
Agronomist
Nematologist
Entomologist
Geneticist and Radiogeneticist
Nutritionist — Plant, Animal, and Human
Botanist
Biologist and Microbiologist
Plant Physiologist
Soil Scientist
Plant Ecologist

My own research into making this list has failed to find a specialty devoted to an obvious bean problem — stress. You know, beans with too little water or too much crowding are under stress, so should not there be a Psychobotanist?

I do not pose as an expert, to give the one perfect way to select, plant, grow, and harvest beans. Rather, my purpose is to present an experienced grower's overview of the many methods and practices which are, or promise to be, successful in growing a crop of beans.

Screening all of the material for this book has been the equivalent of a four-year degree major, with no filler courses. One of the pleasant surprises was finding that a grower's guide, published when I was a teenaged market-garden worker, recommended a program for bean culture which would not lead one astray even today.

Let me hope that my suggestions here will aid and abet bean growers for as many years. It is my pleasure to try.

SOILS

Planting hills of beans or hills of corn may not be the best technique today, now that the hand hoe has been hung up and the power tool has taken over. But the soil requirements are still the same. Friable, "humusy," slightly acid soil, well drained, not too rich in nitrogen, and in full sun, seems to be right for beans.

At least sixteen elements are necessary for the growth of green plants — carbon (C), hydrogen (H), oxygen (O), nitrogen (N), phosphorus (P), sulfur (S), potassium (K), calcium (Ca), magnesium (Mg), iron (Fe), manganese (Mn), zinc (Zn), copper (Cu), molybdenum (Mo), boron (B), and chlorine (Cl). Do not worry about whether your soil has all of them; proceed with the thought that they are all present.

When a grower gets the urge to grow beans, regardless of the time of year, action should begin to bring the plot toward the ideal condition. If the "right" soil does not seem to be available, the gardener should do as bean growers have done for generations — try anyway. Learn by doing. Beans are not an income or survival crop and are quite tough, so be bold. Improve the land bit by bit, year by year.

Few of us can hope for that perfect soil type, so we simply contend with our available site. Tilling in animal manure, compost material, even rotted sawdust or any woody waste will open up heavy soil, making it easier to work, drain better, and give better aeration in the root zone. Light sandy soils benefit greatly from organic matter in an opposite way. Humus holds moisture, enhances the nutrient supply, and gives "body" to the soil.

The action to be stressed here is tilling. Spade, plow, rotary tiller — they simply are tools for stirring up the soil. "Working in" organic matter or "working up" the land is a mixing process which loosens the top 6 or 8 inches of the soil. The method of doing this is the grower's choice. Loose soil makes it easier to get compost in to rot, fertilizer down below plant roots, and seeds in a good sprouting position. Tilling will also bury some weeds and get planted seeds off to a good start before weeds re-start. Turn in organic matter, especially green material, as many weeks ahead of planting as possible to forestall possible trouble with root rot organisms.

Start with an easy step: checking for acidity. Most of the eastern half of the country is chronically acid, so "sweetening" or alkalinizing the soil is a regular procedure. In the West, oversweet soil is not unusual and can cause a "locking up" of some soil nutrients.

The pH scale, a measure of soil acidity or alkalinity, reads from 0 to 14. Beans do best in soils with a pH range of 6.0 to 7.0, but will produce a crop in soil with a pH level as low as 5.5. Problems do not develop until the pH goes below 5.2 or over 7.0.

To find the pH of a plot of land, one should take advantage of the free service offered by state and federal agencies. They will suggest the correct procedure, and their results will often include information on the amount of humus buildup needed to give the soil a better physical tilth.

It is an accepted fact that when bean soil tests show a pH level below 6.5, then sweetening or liming is desirable. Liming may also supply calcium and magnesium while correcting acidity. Correcting alkaline soils, or acidifying them, can be a complicated procedure involving, oftentimes, an excess of many salts. The recommended procedure is to consult with the nearby Soil Conservation Service or county extension station.

Liming usually infers the use of finely ground limestone, which is readily available. Other products are dolomitic limestone, basic slag, marl, hydrated lime, and even oyster shells in some areas.

The amount of ground limestone needed will vary, according to the type of soil and the region. Cool temperate regions require a bit more for the same result in the same soil type. For example, to raise the pH from 5.5 to 6.5 in sandy loam in our northern and central states requires about 55 pounds of fine limestone per 1000 square feet; in clay loam, 120 pounds. Southern coastal-state soils need 35 pounds and 100 pounds in the same soil types.

In the days before nitrogen fertilizers were common, beans were one of the legumes grown to improve the soil through their nitrogen-fixing ability. Beans needed a well-limed soil to accomplish this, and considerable importance was placed on the liming of bean soils. Still, much of this acreage could use more limestone even today, perhaps due to a gradual, natural acidification process.

Ground limestone can be applied at any time of the year, but the usual practice is to incorporate it in the fall after crop debris has been removed. It is not fast acting, but the fine dust has been known to help when spread beside rows of beans growing in acid soils.

FERTILIZING

At the time limestone is added or the plot is being otherwise cleared and worked up for beans, it is a good idea to also add humus material. Rotting and disintegrating plant material, animal manure, etc., provide a soil texture that makes rooting easier, holds moisture, and permits easier cultivation. Soil is more productive if not too heavy, or too sandy either. Both conditions are corrected easily by adding some form of humus.

Humus, which is organic matter broken down by soil organisms into minute particles, should be discussed here because of its relationship to nutrients. As organic matter breaks down, nutrients are both formed and changed to water-soluble elements which can be utilized by plant rootlets. Not all of the potential in the organic matter is released, however. Some elements become "locked up" in forms not available to plants. Nitrogen, phosphorus, and sulfur are three of the most readily available, but seldom in really significant amounts. Soil organic material does contain 5 or 6 percent nitrogen, but almost none gets to crop plants. It is consumed by the bacteria which break down the organic matter.

Remember that organic material should be regularly re-supplied to maintain good soil health. One of the lessons being constantly taught and least remembered among gardeners is that of the transient nature of some of the nutrients, notably nitrogen, which leaches easily. Nitrogen is grabbed by every living organism, from soil bacteria to bean plants. Its absence was noted even in early agriculture, hence the practice of using bean plants and other legumes to add it to the soil.

Even before bacteriology became a science, a French chemist and farmer, J. B. Boussingault, in Alsace, discovered that free nitrogen in the air is changed into compounds suitable for plant growth (legumes) by something that is alive in the soil. Not until fifty years later were the bacteria, later known as rhizobia, isolated. That was a discovery of great importance, and bean growers are the beneficiaries. The rhizobia probably furnish most of the nitrogen that goes from the air nitrogen-to-soil cycle, but they cannot do this unless they are living in the bean or legume nodule, which is formed on the roots. It is estimated that 50 to 150 pounds of nitrogen can be fixed by nodulated legumes on an acre each year.

The right kind of bacteria must be present for each legume plant, and to make sure, one must use packaged inoculants specific for the legume — in this case, garden or field beans. The earlier practice of growing beans on the same land over years did increase the correct rhizobia strains, but rotation to eliminate diseases is now preferred, so packaged inoculants are a must. (They are sold by most major mail-order seed companies and at garden supply centers. Buy only what you need for one season, since the inoculant will not keep for more than one year.) Remember also that the nitrogen available to the bean plant does not come until later in the life of the plant, so some nitrogen in fertilizer form is valuable for getting the beans off to a good start.

One reason that commercial growers are not heavy users of inoculants is the fact that the yield increase from their use is not great. Most growers use seed fungicides to obtain good plant stands, and these seem to inhibit the bacteria.

Another point to remember is that the breakdown of large amounts of organic material requires and takes much nitrogen. There is other good evidence that a strong nitrogen supply does increase yields as well as the protein content of the seed. **Lesson:** use both inoculant and nitrogen.

Actually, beans *can* be grown without any soil at all. Nourish-

ment of the plant by all the sixteen or so necessary elements mentioned earlier takes place through rootlets which can absorb only liquids. The elements are dissolved in liquid water — they are in solution. Plant roots search widely in moist soil to reach the tiny droplets containing all the nourishment, and in the process they build their own cellulose material which becomes the stems or supports for the plant.

In hydroponic culture an inert material such as sand supplies the same base for plant roots to expand in, but the liquid solutions upon which plants feed are the same as those in moist soil. One difference would be that the rhizobia bacteria in soil would not be present at the roots of beans in hydroponic culture. The bacterium named rhizobia is naturally present or may be introduced into bean soil, and when established in the roots can extract air nitrogen for itself and the plant. This "free" nitrogen is not needed, of course, in a complete hydro solution, or in soil which contains enough from fertilizers.

Let us assume that trace elements in the soil are sufficient for good bean growth, otherwise, a whole battery of expensive and perhaps even worthless tests will have to be made. This is not practical for small acreage gardeners. Serious deficiencies in manganese, boron, and other elements can be detected during growth, and remedial action taken then.

This leaves the other two primary nutrients to consider as additions to the soil for a perfect seedbed — phosphoric acid and potash. We think of these "big three" elements as valuable in this way: nitrogen (N) for leafy growth, phosphoric acid (P) for flowering and production, and potash (K) for root growth. Bone meal and rock phosphates supply the phosphoric acid needs of the garden, and wood ashes are the oldest source of potash.

However, the sensible way to fertilize the bean plot is to use a prepared fertilizer, which will have a known nutrient composition printed on the container. Knowing the ingredients and the amounts used will give a base for later adjustments. Also, because the plant can absorb only the dissolved basic elements, it matters little what the source of it is. To quote a soil chemist: "Chemical and biological processes occurring in the soil solution and at the interfaces with solid soil particles create the ions necessary for plant nutrition . . . Biologic processes are required to convert organic sources to an ionic state that is available to plants."

The same government extension stations that offer acidity tests of your soil samples will make free analyses of the nutrients in the soil and give a suggested formula, with amounts to add for bringing the garden to a good nutritional level.

There is good reason for checking up on this matter, and it stems from the fact that some soils in some areas of the country will be found to need no potash, or no phosphoric acid, or twice as much nitrogen, or a combination of all of the above. The suggestions given here, and further on, will be good ones for most of the eastern half of the country, but the extra surety of an aggie consultant's expertise is worth having.

Bean fertilizers will generally be in the range of 5(N)-8(P)-7(K), or perhaps 5-10-5. Stronger units will be 10-16-14, or 10-20-10, which means that less "filler" is used and your applications can be halved. For example, suppose the bag lists a 5-10-5. That is *parts* of the whole bag amount, or twenty parts (or percent), and the remaining bag ingredients will be eighty parts of nothing but sand. Filler is not bad, but actually useful for making the spreading of fertilizer more accurate and uniform.

When a gardener concentrates on beans as I do, or any single crop on a limited area, the rotation of that crop through various areas is important for several reasons. One is disease control. Some diseases like to stay in a soil where the host plant keeps it alive. Without the host plant there is less disease. Another reason for rotation is to keep down possible buildup of nutrients, notably phosphoric acid. Excess amounts of some fertilizer elements are detrimental to plants, and a rotation of both plant food and plant is good practice.

If the grower is bold, and has a disinclination to be precise and "scientific," he has company, and possibly will grow great beans. After all, beans are adaptable, and the soil perhaps has the richness and tilth they need. If there are any doubts in the minds of the bold ones, here is a suggested fertilizer application which will work as well on most lands today as it did when first suggested fifty years ago — medium soils: 2-8-6; sandy soils: 5-10-6. Apply at 800 pounds per acre, or 40 pounds on a plot 30 feet by 70 feet. It should be banded (laid down in a continuous row) 2 inches below the level of the seeds and 2 inches to each side, never right under the seeds.

Modern planting machines make banding easy, but home gardeners will find it more difficult. Remember that the first young bean rootlets should not get "burned" by contact with concentrated fertilizer nutrients. A 2-inch distance beside and below the seed will let the young roots "seek" nourishment in a zone where the plant food will have been dissolved and diluted by more moisture. The earliest rootlets seem to push directly downward.

Old-timers used to plow furrows, then run in the fertilizer, incorporate it by dragging chains down the bottom of the furrow, drop in

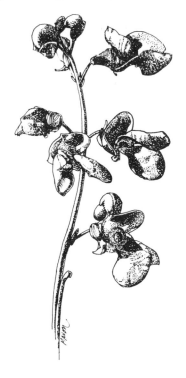

the seeds of beans, corn, or cut potato, then cover them with a hand or horse hoe. Probably there was some "burning" that way, which may explain why not too many still use the method.

One further reminder: watch growth and side-dress as necessary with plant food by scratching it in lightly 4 inches away from the stems. Leaching during rainy periods may necessitate this.

PLANTING

There are three types of growth among beans — bush or dwarf, twining, and climbing or pole types — and row width and seed spacing should be planned to accommodate the growth patterns.

In the world of the agronomist, the growth habit of true bush types is determinate, and that of true pole types is indeterminate, while those bush plants having runners with no ability to cling to supports are twining. Runner beans are a separate class of indeterminate plants. They are naturally perennial, but in temperate zones are annual. Some growers in the Southwest have maintained them for several years. Unlike the case with other species, the seed halves, or cotyledons, of runner beans remain underground after germination. Runner beans are more tolerant of cool weather than are other pole beans, and are very popular in England.

Some bush beans will produce mature plants that top out at 12 inches high; others will be a robust 24 inches with almost as great a width. If information on height and width is available at planting time, adjust spacing as necessary to avoid overcrowding.

Intensive studies have been done on bean *row* spacing, with the aim of pushing production per acre to the limit. In commercial dry bean production, very close row spacing brings problems with white mold and other diseases and difficulty with machine access. Small

growers will probably get the same problems with diseases, plus the real headache of weeds. Commercial growers use herbicides for pre-plant weed control, but these are not available to home gardeners.

Studies on bean row spacing have produced some interesting facts that home gardeners should know, and perhaps take into account: rows spaced 18 inches apart have generally outyielded rows 36 inches apart by 33 percent, and rows 12 inches apart have outyielded those 25 inches apart by 50 percent. But rows 10 inches apart have produced only 7 percent more than 30-inch spacing. The studies were carried out over an eleven-year period in Michigan and involved beans allowed to dry on the bush, i.e. field beans. This and other field tests seem to indicate that many types of bush beans can be planted in rows 18 or even 14 inches apart, which will permit access for hand cultivation and picking, yet give some air circulation for plant health.

Double-row planting has a place in the small garden. Having furrows 5 inches apart, with the number of seeds per foot of row length reduced from six to three per foot, seems to lessen the problem of blowdown. Weed control and access for picking snap pods will control the popularity of this system, however.

If beans are known to be twining types, it seems wise to allow room for that habit, and to be thorough in weed control before the spreading plants take over. Brush should be considered and placed after the last cultivation, as the runners grow. Coarse binder twine tied to short stakes and looped under the runners will help keep beans off the wet ground.

As a general guide consider a seed spacing of 2 to 3 inches and a row spacing of your choice, depending on the cultivating method you plan to use. Tests in Michigan have shown that Navy bean plants 3 inches apart in the row were slightly more productive per acre than those 2 inches apart, and far more productive than those placed 8 inches apart.

I like to remind growers that planting beans closely for high production is wasteful of the seeds if they are scarce. According to the Michigan tests, the greatest increase per seed requires plenty of space per plant. For example, on a row 4 feet long, four plants produced 190 pods, while another 4-foot row with twenty-four plants produced a *total* of 267 pods. More production per row, but more seeds to do it. Widely-spaced plants should be supported. Bean roots will spread to fill a row, even those of plants that are widely spaced.

Climbing or pole type beans (indeterminate) are by far the most common of the old-time heirloom sorts — the home-saved, non-commercial beans. Pole beans do have shortcomings which some growers find bothersome. They require supports, which means extra work before and after the crop. They are slower growing, so later to start producing, and are difficult to cultivate except by hand-hoeing.

Happily, most growers ignore such problems and prefer pole varieties for their plus features: most of them produce more, over a longer period, with no lodging, with fewer insect and disease problems, less stoop labor, and longer pods in greater variety.

Planting patterns may vary as widely as imagination permits. A good pole will be 10 to 12 feet long, hardwood (2 inches in diameter at the base), planted 18 to 24 inches deep in the ground. Smooth poles, such as bamboo or iron pipe, should have a length of rough twine wound around them to give the plant a better climbing surface.

Here are some pole-bean planting patterns:

1. Plant a pole, then five or six seeds around it and 6 inches away. When seedlings appear, thin to three or four.

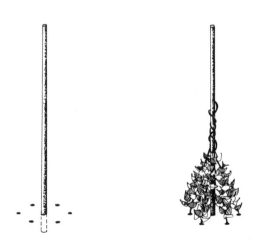

2. Plant a pole, then plant seeds 5 inches apart in a large circle 18 inches or more away from the pole. Later on, tie binder twine in tepee-fashion from high up on the pole to small pegs between plants (a. and b.) or to the plants themselves (c.).

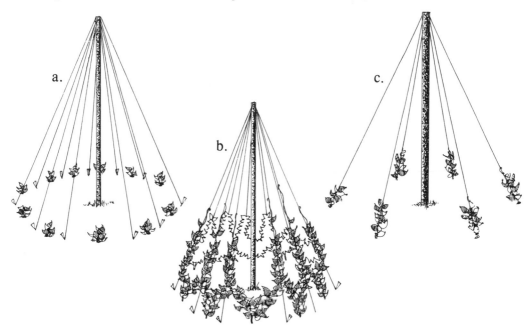

3. Plant a straight row of poles 4 feet apart, then tie plants as in the tepee plan, but in a straight row.

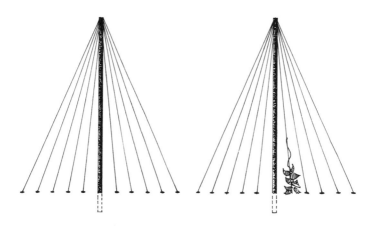

4. Plant from three to six poles in tepee-fashion in circles of various diameters up to 6 feet, allowing three plants per pole. Tie poles near tops.

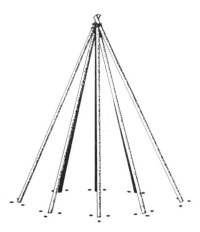

5. Plant single, very strong poles with 2-foot crossbars at 7 to 8 feet up and run twine from the tips of the crossbars to plants in a circle.

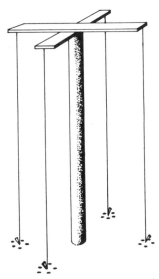

6. Plant strong poles 10 feet apart in straight rows. Fasten 2-foot-long crosspieces or "tees" at 7 feet high on each pole. Run a wire along each end of the crosspieces, and two rows of plants at 5-inch spacing will climb twine to the wires.

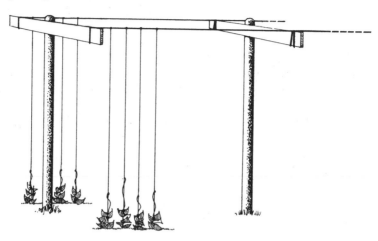

7. Plant old pipe or strong poles in a row at 10-foot intervals. This is a single-row version of the "tee" plan. Run wire along row at arm's reach, and extend twine down to single row of plants spaced 4 inches apart.

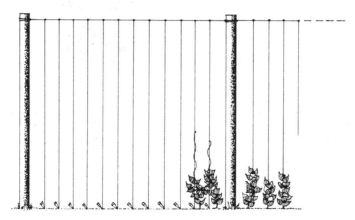

When using either of the last two planting methods, the end poles must be braced or anchored with wire stretched out to stakes. The poles in the rows also need to be strong ones, because the weight of a full crop, plus strong winds, can make a mess of things. I use system # 7, with rows 4 feet apart, and plant rows of bush beans between rows of poles.

Single-row planting is necessary for me to keep separated the nearly two hundred varieties which I renew each year. Beans — all beans! No scientific tests have been made, but it does seem not to harm growth when the twine is loosely tied to the plant, rather than to a low wire along the row. I also use a north-south row plan with the hope that more sun will reach the bush rows. There is no proof that this is best, but it permits free access for note-taking and less chance of "crossover," or tangling across the rows of vigorous plants. Until there is a careful check done, we will not know just how dense a planting of pole beans should be for maximum production per acre.

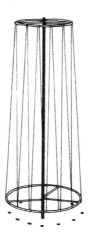 

If you don't have time to build your own support system for pole beans, shop through seed catalogs for a *ready-made model. The one shown above has a wooden base and top, and heavy-duty string "poles."*

Planting beans in corn is a romantic carryover from the native Indian method, and is still practiced. The match-up of strong corn plants and weak bean climbers is critical, because a strong bean plant will fill a stand of corn with a tangle of vines. Picking sweet corn and beans in a jungle is no fun.

Interplant field corn and a climbing dry bean variety for a double crop, but plant the beans a week ahead of the corn so more sun will give the beans a start before the corn shades it.

Widely spaced corn hills and climbing beans have been matched up, and in fact have done well. In Africa and in Brazil, mixed cropping of corn and beans produces higher combined seed yields (as much as 35 percent more) than monoculture at the same planting rates.

Pole bean growers, especially in the northern states, should try this technique to get an early flush of bloom and pods: snip off the tips, or leaders, as the first flowers appear. Secondary growth will follow, with a later crop in good years. Some varieties can be kept within reach by means of this trick.

In arid regions, pole varieties are grown in raised beds, irrigated in the channels, allowed to run unsupported, then the dry beans are harvested like bush varieties. Ground moisture makes this method impractical for most of us, for both pole and twining beans.

The depth to plant beans depends on the size of the seed and the moisture in and type of soil. I plant small beans 1 inch deep, and large ones 2 inches deep. The important thing is that once the seeds are planted, the soil should receive a continuous supply of moisture to prevent it from crusting over. It is not necessary to plant seeds "eye down." Nature will prevail if given a chance. Any attempt to gain a fast start by planting beans, at any depth, in soil that is too cold is taking a chance. The optimum soil temperature for bush beans is 60°F.; for pole and lima beans, 65°F.

Planting time. It is not good practice to try to state calendar dates for planting beans. Even in the small area of the New England states, the natural advance of warm spring weather will produce bean-planting times three weeks apart in a distance of less than seventy-five miles. Some sections of the South can sow and reap two crops of beans in twelve months, sowing in February and in August. In many of our northern states a good bean-starting time can be as late as June 15. Consult your nearest good-hearted, bean-growing gardener!

Many tests have proven that beans planted too early will rot in the ground, or develop root rot if they do sprout, and will grow but slowly in cold soil. Some seed companies offer beans which have been coated with a fungicide (captan), which is claimed to protect against the problems of cold soil. No specifics are given as to how low a soil temperature these treated beans will tolerate. Commercial growers, as well as eager home growers, feel that early starts are necessary to permit long-season varieties to mature, or to capture an early market. The earliest pick of snap beans is as much a gardener's challenge as the first peas or first corn.

It is here that the small gardener can beat the "big guys" by pre-sprouting, or pre-plant sprouting, which is simply a way of getting a 100 percent stand growing early in the season. Not only does this method overcome the hazards of cold soil, it provides a way to get a full plant population from old and low-germination bean seeds.

The pre-sprouting method is simple: wash and then soak beans in plenty of tepid water for a few hours until they swell. Rinse again gently and place the seeds in a single layer on moist paper towels. Cover with another moist paper towel, in contact with the seed. Enclose in plastic or cover to preclude evaporative cooling and keep at 70° to 80°F.

Examine the seeds daily. When the sprouted rudimentary root (or radicle) first appears, plant the seed carefully, firming the soil lightly over it. It is very important that the sprouted seed be planted gently, in a bed that is ready. The nearer the outside soil temperature is to the ideal 60° to 65°F., the less shock and setback. Keep the soil moist for a few days after planting. Sprinkling the planted row will provide soil moisture until the roots take hold. If outside garden conditions are not right when the sprouted beans are ready for planting, they may be held over by sowing them individually in peat pots, which can be planted later, without disturbing the roots.

I feel it is important to mention right here that a lot of energy devoted to hurry-up tricks can be wasted. There is a rule of nature which sees to it that warm-weather plants grow best in weather that is warm. It is no news to old gardeners that beans, planted early, in open ground, will often not show flowers one day earlier than beans planted two weeks later, when conditions were better. The early-planted beans "just sit there," as they say. Sometimes the take-a-chance planting *will* work out well, so, have fun, but be aware!

Cloches. Another way to rush the season, on beans and other crops as well, is to erect temporary tiny greenhouses over single rows. They are called *cloches* in France and England where the system is very popular. They help to warm the soil under them and the plants inside them, giving bean leaves the needed warm temperatures. Cloches are perfect homes for the pre-sprouted beans just described.

And they are simple to make. Use 8-gauge wire in 4-foot lengths and make half-hoops 20 inches wide and 12 inches high, using a template to form them and to keep them uniform. Allow about 5 inches on each leg, then bend a full loop around a nail or pin. This will be the tie point for the nylon cords later. Make enough of these half-hoops to place at 4-foot intervals along a row, pushing the ends in to the tie loops.

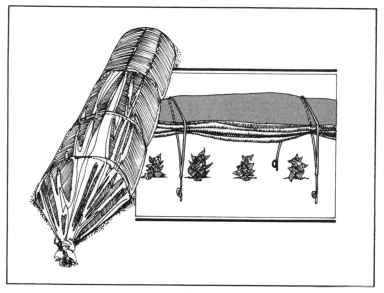

Two views of a cloche: with cover in place (left), and with sides raised for ventilation (right).

Next, stretch a length of 3-foot-wide, 4-mil polyethylene over the row of wires. Tie the end and pull tightly to stakes. Tie lengths of nylon string over the plastic at each hoop to the formed wire loops at ground level. The string will keep the plastic from ballooning in the wind, and on hot days will hold the plastic up off the ground along the row for ventilation. It is a great system, and black plastic used over the soil in addition will warm the ground beneath.

Even in my area (eastern Massachusetts) of usually adequate soil moisture, there are early dry spells as well as midseason ones. The lesson I have learned, and heeded, is that frequent waterings are best. A soaking rain soon after a heavy watering could be the coup de grace for bean roots in any but sandy soil. Beans do not like wet feet.

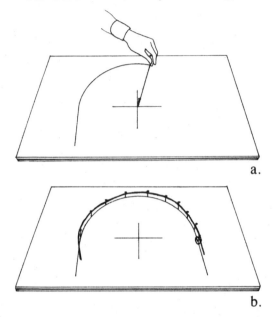

a.

b.

Make a template on a piece of plywood to form supports for a cloche. First, with a string and a pencil, mark off a half-hoop 20 inches wide and 12 inches high (a). Drive nails into the wood along the hoop outline, bend 4-foot lengths of 8-gauge wire into hoops, allow 5 inches on each leg, and bend in full loops around bottom nails to form tie points for cords.

GROWING CARE, DISEASES, AND INSECTS

Although beans have a strong natural will to live and reproduce, there are diseases and insects which seem to exist solely to pester them. Diseases are more to be feared, because they are difficult for non-professionals to recognize and hard to cope with. Many diseases are seed-borne, which raises additional control problems.

A good general rule for small growers is to pull and discard (*not* compost) those plants which are dying or are strongly infected with an obvious problem. Plant injury may not necessarily be caused by disease. Soil deficiency, air pollution, herbicide drift, and even sunscald can discolor and distort bean plant leaves into an appearance of disease. Diseases and the injuries described can be influenced by soil and air temperatures and excess moisture which will vary year to year, so the home gardener need not panic and destroy every suffering bean plant. If the variety involved is a cherished heirloom, seed should be saved by all means, and advice sought. Bean problems are not always malignant or catastrophic. They lurk in soil and seed and invade on the wind. The best defense is attention to details — crop rotation, clean soil, tilling in the fall, and spotting problem plants early.

Here is a list of problems and diseases which will serve as "talking points" in discussions with neighbors: dry root rot, rhizoctonia, halo blight, common fuscous blight, sclerotinia, white mold, anthracnose, rust, bean mosaic virus, bronzing, and sunscald. Look at the bright side — you may never see one of them.

Insects are another ever present problem. In some years, one or another pest may find conditions perfect and overpopulate to cause serious damage. The list of enemies is long: white grubs, wireworms, flea beetles, seedcorn maggots, bean thrips, Mexican bean beetles, cutworms, green cloverworms, bean aphids, potato leafhoppers, tarnished plant bugs, two-spotted mites, and more. Because they seem to occur in unpredictable fashion, large numbers of any insect should be sufficient cause to check with a local agent for control advice, and get a good sprayer ready to work.

The most commonly inquired about pest is the Mexican bean beetle, whose fuzzy yellow young grubs can appear early, and must be searched for often on the undersides of leaves. An organic dust, rotenone or pyrethrum, if applied as soon as the first beetle grubs are seen, will control them, as will malathion or Sevin.

The pest next most frequently asked about seems to be the insect that infests the dry beans in storage. This is the bean weevil, which will be discussed later, under "Storage."

Less than ⅛-inch long, the bean weevil is an active pest that lays eggs on bean pods. Its larvae, as many as a dozen at a time, drill their way into forming seeds and remain there for months until they change to adults and cut their way out again.

BEAN TYPES AND VARIETIES

String beans were correctly named. The pods had, and still have in many varieties, rather strong suture fibers which would not cook tender and had to be "strung" — the tips snapped off and the stringy fibers framing each side removed. The pods were then snapped into cooking lengths. String beans are produced on every type of bean plant — from bush to pole. The pods may be long or short, green, yellow or streaked, round, oval or flat, fibrous or tender. Bean breeders first worked them over to eliminate the strings, and called them stringless. Nowadays, stringless beans are "snap" beans. Only without the strings are they affordable in this costly age.

In some areas of the country, yellow string beans are called wax beans; in other locales they are called butter beans, but not butterbeans, which refers to small limas. So it is with green beans. They are not necessarily green in color — simply fresh, young pods.

Shell beans are immature beans in a certain pod condition, regardless of variety name. The seeds are swollen and the shell has passed the tender, fleshy stage, to the point of being thin and rubbery. It is the stage just before the pod dries out and before the seeds inside have shrunk and become "dry" beans. The term "shell bean" has traditionally been reserved for a class known as horticultural beans, which are superior in the shell stage and have colorful pods. (In some areas, these horticultural beans are called shelly beans or October beans. In California, this class has always been called Cranberry beans; some seed catalogs will list such beans in this way also.)

The quality of the shelled seed is described as farinaceous. Shell beans are eaten as fresh beans, whether they are horticultural, Cranberry, October, kidney, or lima beans. (You can also let any of these beans mature on the vine and use them as dry beans.) Green shell limas are probably more popular than dry limas. Some canners sell a horticultural bean in cans labeled "shell" beans, but these are reconstituted from dry beans.

Lima beans are a distinct class, divided into three types. Small limas are Sieva beans, fat large limas are Potato Limas, and the flat large ones are Flat Limas. When we consider that beans are not too demanding of space in a garden, are adaptable to poor conditions, and produce a green vegetable, a tasty immature seed, and colorful, long-storing edible dry seed, we can appreciate how farm families treasure their heirloom beans — the ones which do well for them and did well for their ancestors.

Heirloom beans have been identified by variety with all sorts of descriptive names, beginning with the direct, simple kidney, in red, black, and white. Some of the old-time beans are used by bean breeders for their adaptability to specific regions of the country and their distinctive cooking qualities. Many are kept viable in government seed-storage facilities, for the express purpose of research. (At the time of this writing, the only "public" system for obtaining heirloom beans is a non-profit organization run by the author of this text. If you're interested in heirloom beans, send $3 to Wanigan Associates, 262 Salem Street, Lynnfield, MA 01940, and obtain a quarterly newsletter on beans, a catalog listing some 500 varieties, and an invitation to select two varieties of your choice.)

There are hundreds of different beans to grow, whether you decide to specialize in heirlooms or to try the latest varieties. A sampling of what is commercially available (including some old-time beans) is shown by the chart that follows.

Keep in mind as you shop for seeds that there are no rules or agreements among seed companies or in government which prohibit or even inhibit seed sellers from making up new names for bean varieties and promoting them. For example, the Yellow Bountiful of yesterday can become the "exclusive" Golden Wonder in a new catalog. This practice breaks no law — just your trust. "Exclusive with us" is a hollow statement made to improve sales and company ego. To date there is no such thing as a hybrid bean.

COMMERCIALLY AVAILABLE BEAN VARIETIES

Variety	Type	Days to Maturity*	Comments	Sources
Alabama Pole #1	pole	75-80	Pods 7" long, stringless when young. Excellent flavor. Plant among corn.	G
Anniversary Gold	bush, yellow	55	Heavy yielder. Tolerates night temperatures above 60° F. Freezes well.	W
Astro	bush	50-52	Pods set high on plant; popular with commercial growers.	H, L, Y
Avalanche	bush	50	Smooth, straight, light green pods.	R, W, Y
Baby Fordhook	bush, lima	70	Curved pods 2"-3" long, 3/4" wide, 3-4 dark green potato-type beans per pod.	A, M
Baby Potato	bush, lima	60	Pods 3" long. Beans freeze well.	U
Bestcrop	bush	52	Pods 6" long, stringless. Endures all kinds of weather.	Y
Beurre de Rocquencourt	bush, yellow	46	Thin, upright pods 6"-7" long. Recommended where nights are cool. Black seed.	J
Black Turtle (Black Turtle Soup)	bush	85-115	Slender pods can be used as snap beans; best used as dry bean in soups, stews, and refried. May produce short runners. Black seed.	J, Y
Black Valentine	bush	48-70	Pods 6½"-7½" long. Use as snap or dry bean; good in soup.	M, P, S, Y
Bluecrop	bush	54-56	Long, round pods. Heavy producer. Freezes well.	F, U, W, Y
Blue Lake	bush	52-61	Round, curved pods 5"-6½" long. Upright growth. Excellent flavor. Seed slow to develop.	A, B, D, E, F, G, H, K, L, N, O, Q, R, S, U, W, Y, Z
Blue Lake Stringless (White Creaseback)	pole	55-65	Round pods 6"-7" long, stringless when young. Use as shell or dry bean.	A, D, H, K, M, N, R, S, T, U, Y, Z
Blue Ribbon	pole	69	Creaseback. Grows to 5'. Tolerates drought.	Y
Bonus Small White	bush	85	Long, narrow pods. Use as snap, shell, or dry bean. Very good baked.	Y

Variety	Type	Days to Maturity*	Comments	Sources
Bountiful	bush	47-50	Thick, flat, pale green pods 6"-7½" long. Good shell bean. Popular all-purpose.	A, H, M, O, Q, B, T, W, Y, Z
Bridgeton	bush, lima	76	Baby lima, averages 4 small green beans per each flat 3" pod. Seed freezes well.	F, G, R, Z
Brittle Wax	bush, yellow	50-52	Round pods 6"-7" long, stringless. Mild flavor. Vigorous. Continues to produce over long period.	A, E, M, T
Buff Valentine			See Contender	
Burpee's Best	pole, lima	92	Pods 4"-5" long, 1¼" broad, 4-5 "potato" limas per pod. Vigorous climber.	A, O
Burpee Golden	pole, yellow	60	Wide, flat pods 5½"-6½" long. Starts like bush bean with pods close to ground; then develops runners.	A, O
Burpee's Improved	bush, lima	70-75	Flat, uniform pods 5" long; 3-5 beans per pod.	A, C, M, O, Q, T, U, Y
Burpee's Tenderpod	bush	50	Brittle, fleshy pods 4½"-5½" long. Long, curving ends.	A, O
California Red Kidney	bush	100	Large red kidney. High yielder.	U
Carolina (Sieva, Southern Running Butterbean)	pole, lima	77-78	Dark green pods 3½" long; 3-4 small, white beans per pod. Use as shell bean. Tolerates cold.	A, G, L, M, O, S, Y, Z
Cascade	bush	54	Dark green pods. Freezes well.	Y
Champagne	pole	62	Flat, stringless pods 8" long. Grows to 20'. Continuous set of pods. Best used as snap bean.	J
Charlevoix	bush	110	Dark red, medium kidney. Provides thick broth in soups, stews, and chili.	J
Cherokee (Valentine Wax)	bush, yellow	50-52	Resembles Black Valentine, but has thick yellow pods 5"-7" long. Dependable in unfavorable weather conditions.	B, C, D, E, F, G, H, L, N, Q, U, W, Y, Z

*To most bean growers maturity means dry seed, which can be used, if desired, for another year's crop. The figures listed here are simply designed as a guideline for reaching the most popular stage for eating, and not, in all cases, the dry bean stage.

Variety	Type	Days to Maturity*	Comments	Sources
Christmas (Giant Florida)	pole, lima	80-85	Pods 5"-6" long. Seed the size of a quarter, light cream with red stripes. Use fresh or dry.	P, R, S, Y
Commodore (Dwarf Kentucky Wonder)	bush	58-65	Round pods 6½"-8" long. Dull reddish-purple seed. Good shell bean. High yielder, recommended for small gardens.	B, G, L, M, R, S, T, Y, Z
Contender (Buff Valentine)	bush	40-55	Slim, round pods 6½"-7" long. Buff seed with brown mottling. Tolerates heat.	B, E, G, H, L, M, N, Q, R, S, U, W, Y, Z
Corn Hill (Speckled Cut Short)	pole	?	Flat pods 4" long. Yields well, good for small gardens. Plant with corn.	T
Crusader	pole	63	Large, fleshy pods.	V, X
Cyrus	bush	50	Slim, round, straight pods. Cook whole.	V
Dade	pole	55	Green pods 7"-8" long, stringless if picked when young. Runners slow to appear. Recommended for hot, humid southern weather; not good for North.	L, Y
Desirée	pole	65	Averages 40 10"-12" pods per plant; small seed and fleshy pods. Excellent flavor. Prefers dry conditions.	V
Dixie White Butterpea	bush, lima	76	Broad, oval pods 3"-4" long. Roundish baby white limas. Will set pods at high temperatures.	G, L, R, S, Y
Dixie Speckled Butterpea	bush, lima	76	Brownish red seed speckled with darker brown. Use as shell or dry bean. Excellent flavor. Allow 90 days for dry bean.	L, R, S, Y
Dwarf Horticultural	bush	55-63	Thick pods 4½"-6" long, splashed red in shell stage. Buff seed with maroon spots. Use as snap, shell, or dry bean. Preferred in succotash in many areas.	A, J, M, N, O, Q, S, T, W
Dwarf Kentucky Wonder			See Commodore	
Earlipol Bean	pole	55	Round pods, almost stringless. Climber, grows well in North.	C

Variety	Type	Days to Maturity*	Comments	Sources
Early Gallatin	bush	53	Round pods 5"-6" long, slightly creaseback.	Y
Earlywax Golden Yellow	bush, yellow	54	Retains quality long after picking. Freezes well.	W, Y
Eastern Butterwax	wax, yellow	44-58	Thick curved, round to slightly oval pods up to 7" long. Freezes well.	F, U, Y
Executive	bush	52-53	Round, dark green pods 5½"-6" long. Light buff seed. Heavy yielder.	W
Evergreen	bush, lima	72	Baby limas borne on compact bushes.	K
Faribo Brittle Wax	bush, wax	45-55	Stringless. High yielder.	C
Flair	bush	42-45	Oval-to-round pods 5" long. Plant for fall crop.	X
Florida Speckled Bean (Speckled Butterbean, Calico Pole)	pole, lima	78-85	Clusters of pods 3"-3½" long. Whitish-buff seed, speckled maroon. Excellent shell bean. Bears well in hot weather.	D, E, G, L, N, S, T, Y, Z
Fordhook 242	bush, lima	72-85	Thick, straight pods 3"-5" long. Plump greenish-white seed. Very popular. Plants bear until frost. Tastes like chestnuts. Seed freezes well.	A, C, D, E, F, G, H, J, K, L, M, O, Q, R, S, T, U, W, Z
French Horticultural	bush	64-68	Pods 6"-8" long, yellow splashed with bright red. Produces runners. Usually not staked. Use as shell or dry bean. Freezes well.	E, F, H, U, Y
Gardengreen	bush	55	Round, dark green pods 5"-5½" long. Produces over long period.	E
Gator Green	bush	53-55	Slender pods 6"-7" long. Tolerates poor weather. Freezes well.	W, Y
Genuine Cornfield	pole	71-81	Round, light green pods 6" long. Seed buff with brown stripes. Use as snap or shell bean.	G, Y
Giant Florida			See Christmas Lima	

*To most bean growers maturity means dry seed, which can be used, if desired, for another year's crop. The figures listed here are simply designed as a guideline for reaching the most popular stage for eating, and not, in all cases, the dry bean stage.

Variety	Type	Days to Maturity*	Comments	Sources
Giant Stringless	bush	45-57	Round, meaty, brittle, pointed pods 6"-7" long. Freezes well.	M, T
Gina	bush	60	Pods 7"-8" long. Produces 2' horizontal shoots. Excellent flavor. Prolonged yields.	X
Gitana	bush	45-55	Slender, stringless pods. Extended tips.	X
Gold Crop	bush, yellow	45-54	Round, stringless, deep yellow pods 6"-6½" long, borne high off the ground. 1974 All-America winner.	A, C, H, K, N, R, U, V, W
Golden Wax Improved			See Top Notch	
Great Northern White	bush	85-90	Larger than Navy beans. Hardier, earlier, more productive than Navy. Use in 65 days as shell bean. Excellent flavor.	B, C, D, E, N, Q, T, Y
Greencrop	bush	42-51	Flat, dark green pods 6"-8" long. Bred for North; does not perform well in heat or drought.	A, C, D, J, O, U, W, Y
Green Isle	bush	45-50	Round, dark green pods 6"-8" long borne well off ground. Heavy yielder. Freezes well.	C
Greenpak	bush	60	Bush Blue Lake variety. Pods 5½"-6" long. Sweet flavor.	W
Green Perfection	bush	58	Smooth, meaty, round pods 6" long, borne off ground.	W
Green Ruler	bush	51	Improved Bush Romano type. Long, flat pods. Heavy yielder. Good beany flavor.	F
Greensleeves	bush	56	Round, thick, meaty pods borne well off ground. Freezes well.	A
Harvester	bush	50-60	Straight, round pods 5"-6" long. Resists rust and root rot.	H, L, M, S, U, Y
Henderson (Bush Butterbean)	bush, lima	65-81	Baby lima. Pods 3" long, 3-5 seeds per pod. Light green in shell stage; white when dry. Excellent flavor.	A, B, D, G, H, J, K, L, M, N, O, Q, R, S, T, W, Y, Z
Honey Gold	bush, wax	40	Straight, round, honey-colored pods 5½" long.	U

Variety	Type	Days to Maturity*	Comments	Sources
Horsehead	bush	100	Dark red mottled seed. Use as kidney bean.	X
Iprin	bush	50	Very slender pods. Good flavor. Can be cooked whole.	V
Jackson Wonder (Old Florida Bush, Speckled Butterbean, Calico)	bush, lima	65-83	Broad, flat pods 3" long. Use fresh or dry. Small seed, gray in shell stage, dark brown mottled with purple in dry stage. Semi-runner. Resists drought.	G, J, L, M, Q, R, S, T, Y, Z
Jacob's Cattle (Trout)	bush	85-88	Use as shell or dry bean. Kidney-shaped white seed with dark red speckles.	J, Y
Joy	bush	54	Straight, round, waxy pods. Seed slow to develop, forms meaty beans. Freezes well.	M
Kelvedon Marvel	pole	65	Long straight pods. Heavy yielder. Produces runners.	X
Kentucky North	pole	64	Long, round, dark green pods.	J
Kentucky Wonder (Old Homestead Pole, Texas Pole)	pole	58-72	Thick, oval, curved pods, 6"-10" long. Very popular. Use fresh or dry. Buff-brown seed. Tender, slightly stringy. Vigorous climber.	A, B, C, E, G, H, I, J, L, M, N, O, P, Q, R, S, T, U, W, Y, Z
Kentucky Wonder Wax	pole, yellow	65-68	Transparent yellow pods 6"-9" long. Fleshy, flat-to-moderate oval, with strings. Light chocolate-brown seed. Strong climber. Use fresh.	A, D, K, M, Q, T, Y
Kentucky Wonder, White Seeded (Burger Stringless)	pole	60-66	Oval, round, stringless pods 8"-9" long. White seed. Use as shell or dry bean. Vigorous vines.	F, M, S
Kinghorn	wax, yellow	45-54	Straight, round brittle pods 5"-7" long.	E, H, K, N, O, U, W, Y
King Horticultural (Worcester)	pole	75	Broad stringless pods 6" long. Not all are straight. Bright red markings on pods at shell stage. Use as snap or shell bean.	F

*To most bean growers maturity means dry seed, which can be used, if desired, for another year's crop. The figures listed here are simply designed as a guideline for reaching the most popular stage for eating, and not, in all cases, the dry bean stage.

Variety	Type	Days to Maturity*	Comments	Sources
King of the Garden	pole, lima	85-90	Flat green pods 5″-8″ long; 4-6 greenish-white seeds per pod. Honey-like taste. Vigorous climber, needs good support. Use fresh or dry.	A, B, E, G, H, L, M, N, O, R, T, U, W, Y, Z
Landreth's	bush	50-52	Slim, round, dark green pods. Heavy yielder. Tolerates drought.	E, M
Largo	pole	?	Long straight pods. Flavor of dwarf French beans. Use as snap bean. Climbs.	X
Light Red Kidney	bush	?	Broad, thick, fleshy pods. Very sweet. Use dry for soups or main dishes.	J
Lika Lake	bush	56	Flavor and quality of Blue Lake. Straight, dark green pods 4″ long. Seed develops slowly. Meaty snap bean.	O
Limelight	bush	38	Thick flat pods, sweet flavor. Use as snap or dry bean.	V
Loch Ness	bush, yellow	52	Round pods, upright growth. Juicy with good flavor. Seed develops slowly.	V
Louisiana Purple Pod	pole	67	Purple pods 7½″ long, turn green when cooked. Almost stringless when young. Use as snap bean.	S
Low's Champion	bush	65	Long, flat pods with 5-6 purplish seeds. Premier snap and shell bean.	J, Y
Majestic	wax, yellow	58	Long, fleshy pods. Seed slow to develop.	W
Masterpiece	bush	50-60	Long pods, high yielder.	X
McCaslan	pole	?	Similar to Kentucky Wonder but has longer pods, up to 8″, flat and slightly curved. Use as snap, shell, or dry bean. White seed.	G, L, T, Y
Missouri Wonder	pole	65	Long, round pods. Mottled tan seed with brown stripes. Plant with corn. Use as snap bean.	T, Y

Variety	Type	Days to Maturity*	Comments	Sources
Michilite	bush	85-90	Improved Navy bean from Michigan Ag. Station. Small, oval, white seed. Use as shell or dry bean.	C
Montezuma Red	bush	95	Good dry bean. Excellent baked.	P
Moongold	bush, yellow	55	Pods 5"-5½" long. Heavy yielder. Excellent flavor.	D
Morse's Pole #191	pole	63	Type of White Seeded Kentucky Wonder. Dark green oval pods 8"-9" long. Pick when young and stringless.	L
Navy (Pea Bean)	bush	85-95	Small, white, oval beans; use dry for soups and baking.	D, E, N, T, Y
New Sanilac Field	bush	95	Navy-type. No runners. Glossy white beans good for soups and baking.	K
Old Dutch Half Runner	bush	57	Round pods. Remains tender on vine in snap stage. Bears longer if staked.	R
Oregon Blue Lake	pole	60	Oval to round pods 6½" long. Use fresh or dry. Freezes well.	P
Oregon Lake	bush	55-58	Dark green pods 6½" long. Use as snap bean.	P, Y
Oregon Giant Paul Bunyan	pole	63-68	Uniform, thick-meated pods 1' or longer, 4-5 make a meal. Excellent flavor. Vigorous climber. Use as shell bean.	P, Y
Painted Lady	pole	?	Smaller pods than Scarlet Runner. Use as shell or dry bean.	I
Parker Half Runner	bush	65	Oval pods 5" long. White seed. Use dry for baking.	M
Pencil Pod Wax (Butter Bean)	bush, yellow	45-53	Straight, round, brittle, fleshy pods 5"-7" long. Black seed.	A, B, C, H, K, M, N, O, Q, T, U, X, Y
Pink Bean	bush	60	Long narrow pods. Medium pink beans hold shape when cooked. Use as shell bean, or in 85 days as dry bean. Best in chili, soup, or baked.	Y

*To most bean growers maturity means dry seed, which can be used, if desired, for another year's crop. The figures listed here are simply designed as a guideline for reaching the most popular stage for eating, and not, in all cases, the dry bean stage.

Variety	Type	Days to Maturity*	Comments	Sources
Pink Half Runner (Peanut)	bush	50-60	Pods 4½" long. Pink seed. Use as snap or dry bean. Performs well in dry climates.	M, Y
Pinquito	bush	90	Tiny pink beans, hold together well when cooked. Use dry for baking. Excellent in Portuguese dishes.	P
Pinto	pole	90-96	Good dry bean, popular in Mexican dishes. Can be used as snap bean. Buff seed speckled with brown. May run.	A, D, E, J, M, R, Y
The Prince	bush	50-60	Long, dark green pods, almost stringless. Excellent flavor.	X
Prizetaker	pole, lima	90	One of largest beans. Pods 6" long, 1½" wide, with 3-5 giant seeds per pod.	A, O
Prizewinner	pole	60-65	Large, fleshy pods. Good flavor. Produces runners.	X
Processor	bush	55	Round to oval pods. Use as snap bean. Flavorful when mature.	X, Y
Prolific	bush, lima	68	White seed, slightly larger than Henderson.	Z
Provider	bush	50	Straight, round pods 5"-6" long. Buttery flavor. Purple seed. Use as snap bean.	F, J, U, W, Y, Z
Puregold	bush, yellow	59	Slim, brittle, round pods 5½" long. White seed with brown eye. All-America Winner. Will not set fruit where nights are over 60° F. Good for fall crop.	C
Purple Pod	pole	62-65	Deep purple pods turn light green when cooked. Use as snap bean when young.	D, E, H, M, Y
Ranier	bush	50	Round pods 5"-6" long. Can be used raw; requires very little cooking.	B, Y
Rattlesnake	pole	86	Dark green purple-streaked pods 7" long. Climbs to 10'.	Y
Red Kidney	bush	95-112	Tough, flat dark green pods 6" long. Reddish-brown, kidney-shaped seed. Use as dry bean in Mexican dishes, baked, and in soups.	A, B, C, D, E, J, M, N, O, P, Q, R, S, T, Y

Variety	Type	Days to Maturity*	Comments	Sources
Redkloud	bush	100	Early Red Kidney. Good for baking.	F
Red Mexican	bush	85	Good baking bean; does not become mushy when cooked.	Y
Remus	bush	48	Straight, round, dark green pods up to 10″ long, borne well above ground. Good salad bean.	V
Richgreen	bush	56	Rich green pods 5″-6″ long. Retains color when cooked. Use as snap bean.	B, O
Roma	bush	59	Similar to Romano pole. Flat pods 4½″ long. Use as snap bean.	A, G, H, O, P, R, Z
Romano (Italian Bush)	bush	50-58	Thick flat pods 5½″-6″ long. Bears for 6-8 weeks. Use as snap or green shell bean.	D, E, K, L, M, U, W, Y
Romano (Italian Pole)	pole	64-70	Broad flat pods 5″ long, show tinge of red streaking at shell stage. Excellent flavor. Use as snap or shell bean. Pods freeze well. Weak climber.	A, B, D, F, H, M, N, O, P, Q, T, Y
Royal Burgundy	bush	50-55	Curved, dark purple pods 5″-6″ long turn green in 2 minutes when cooked. Freezes well. Use as snap bean.	A, K, O, S, U, V
Royalty	bush	51-55	Bright purple curved pods 5″ long. Produces short runners. Use as snap bean.	C, D, E, G, J, M, O, P, R, T, Y
Salem Bush Blue Lake	bush	50	Flavor of Blue Lake pole bean. Meaty pods 5½″ long.	C
Sanilac Navy	bush	95	Developed by Michigan Ag. Station. Pods borne high on plant. Glossy white seed. Use as dry bean.	Q
Scarlet Runner	pole	65-90	Highly ornamental. Pods 6″ long are green at first, turn wine color at maturity. Scarlet flowers appear in 65 days. Vines climb to 10′. Large reddish-brown seed, mottled black. Use young pods as snap, older ones as shell and dry bean.	F, H, I, M, Q, T, U, Y

*To most bean growers maturity means dry seed, which can be used, if desired, for another year's crop. The figures listed here are simply designed as a guideline for reaching the most popular stage for eating, and not, in all cases, the dry bean stage.

Variety	Type	Days to Maturity*	Comments	Sources
Scotia (Striped Creaseback)	pole	72	Light green pods 6″ long, ½″ wide, with purple spots. Use as snap bean.	T
Seafarer	bush	90	Navy bean. Small, round, shiny white seed. Use dry, baked, or in soup.	J, M
Selka Improved	pole	60	Fragrant pods, flat, 10″-12″ long. Stays tender on vine.	V
Selma Zebra	pole	55	Light green pods striped with blue. Beany taste, distinctive aroma.	R
Sieva			See Carolina	
Slenderette	bush	53	Shiny, dark green pods 5″ long. Small white seed, slow to develop. Use as snap bean.	N
Slender White	bush	56	Dark green pods 5″-6″ long. Used extensively for commercial freezing.	H, Y
Slim Green	bush	60	Thin pods 6″ long. Produces lush growth. Use as snap bean.	H, Y
Soldier	bush	85-89	Large, white, kidney-shaped bean with maroon figure of soldier on eye. Use dry for baking and stew.	J, Y
Spartan Arrow	bush	42-51	Meaty, oval pods 5½″-6½″ long.	C, D, F, R, U
Speckled Cranberry (Bird Egg Bean, London Horticultural)	pole	65-69	Old-time variety. Slender oval pods 8″-9″ long. Flesh-colored seed with red spots. Use as shell or dry bean.	T, Y
Speckled Pole Lima	pole, lima	85	Pods green at first, splashed with red at maturity. Nutty flavor.	N
Speculator	bush	45	Straight dark green pods borne well off ground.	U
Spring Green	bush	41	Round, dark green pods 4″-6″ long. Use as snap bean.	U
Sprite	bush	54	Slim, straight, upright pods. Use as snap bean.	Y

Variety	Type	Days to Maturity*	Comments	Sources
State Half Runner	bush	60	White Half Runner variety. Oval, curved pods. Strings must be removed before cooking.	G, M, R
Streamline	pole	65	Pods 15"-18" long. Produces runners.	V
Stringless Green Pod (Burpee's Stringless)	bush	45-53	Round, dark green pods 5"-7" long. Excellent flavor. Light coffee-brown seed. Use as snap bean.	A, B, C, D, I, M, O, Q, T, Y
Stringless Red Valentine	bush	47	Crisp, curved, crease-back pods 4½" long. Use as snap bean.	M
Striped Half Runner	bush	60	Pods 5" long, stringy at maturity. Use as snap bean when young, or later as shell bean.	M
Sulphur (Golden Cranberry)	bush	55	Round pods 5" long. Popular in high elevation areas of North Carolina, Virginia, and Tennessee. Distinctive flavor when baked.	M
Sungold	bush, yellow	45-56	Straight, round, slim, bright yellow pods 5"-6" long. Seeds develop slowly. Compact vines.	F, J
Surecrop Stringless Wax (Yellow Bountiful)	bush, yellow	52	Straight-to-slightly-curved flat pods 6"-6½" long. Stringless at all stages.	A, M, O
Taylor's Horticultural (Shelly Bean)	bush	60-64	Cranberry-type bean. Light green, mostly straight pods 5½"-6½" long, turn cream splashed with red at maturity. Use as snap or green shell bean.	G, L, M, Z
Tempest	bush	100	Use as dry bean. Harvest when pods are full and begin to turn yellow.	X
Tendercrop	bush	46-54	Meaty, brittle pods 5"-5½" long with pointed tips. Holds shape on vine and retains quality after picking. Mottled purple seed. Use as snap bean.	A, D, E, H, P, R, U, Y
Tenderette	bush	55	Dark green pods 4"-6" long. Performs well in hot weather.	D, E, G, H, M, Q, R, T, Z

*To most bean growers maturity means dry seed, which can be used, if desired, for another year's crop. The figures listed here are simply designed as a guideline for reaching the most popular stage for eating, and not, in all cases, the dry bean stage.

Variety	Type	Days to Maturity*	Comments	Sources
Tendergreen	bush	45-57	Brittle round pods 6″ long. Mottled brown seed. Erect vines. Performs well in hot weather.	A, B, C, D, E, G, I, K, L, M, N, O, Q, R, S, T, U, V, Y
Tennessee Green Pod	bush	50	Flat, broad pods 6″-8″ long, become stringy with age. Distinctive flavor. Yellowish-brown seed. Not recommended for North.	J, L, M, S, Y, Z
Thaxter	bush, lima	67	Developed by USDA. Pods 3″ long. Similar to Thorogreen.	H
Thorogreen (Green Seeded Henderson, Cangreen)	bush, lima	65-72	Pods 3″ long, borne high above ground. Small, flat, green baby limas, hold color when dry.	C, G, M, N, R, S, Y, Z
Tidal Wave	bush	54	Dark green pods 6″ long. White seed. Use as snap bean.	Y
Topcrop	bush	43-52	Developed by USDA. Round, straight pods 5½″-7″ long. Brown seed with buff splashing. One of earliest disease-resistant varieties.	A, B, C, D, E, G, H, K, L, M, N, Q, R, T, U, W, Y, Z
Top Notch (Golden Wax Improved)	bush, yellow	50	Wide pods 4″ long.	A, B, C, D, E, H, M, P, R, S, T, Y
Vermont Cranberry	bush	60	Pods easy to shell. Produces 5-6 seeds per pod.	Y
Vermont Cranberry Pole	pole	60	Climber. Use as shell bean or let dry on vine and use for baking.	Y
Violet Podded Stringless	bush	63	Deep blue pods 8″ long turn green after boiling for 2 minutes. Excellent flavor.	V
Wade	bush	50-58	Straight pods 6″-6½″ long. Pick early pods to use as snap beans; leave others on for shell beans.	C, G, Y, Z
White Half Runner (Mississippi Skip, Mountaineer)	bush	60	Light green pods 4″-5″ long, stringless if picked young. Use as snap, shell, or dry bean. Produces runners.	A, J, L, M, T, Y, Z

Variety	Type	Days to Maturity*	Comments	Sources
White Kidney	bush	100	Long pods. Produces vigorous vines. Use dry for baking.	F, E, M, Y
White Marrowfat	bush	100	Pods 5" long, 5-6 seeds per pod. Use as dry bean for baking.	A, M, T, U, Y
Wonder	bush, lima	75	Produces 4-6 large seeds per pod. Requires no staking.	K
White Butterpea	bush, lima	72	High yielder. Rich flavor. Seed freezes well.	Z
White Flageolet	bush	100	Popular in south of France for salads and casseroles. Use white seed as shell or dry bean.	Y
White Mexican	bush	90-95	Navy bean. Large upright pods. Medium-white, almost round beans.	T
Woods Prolific	bush, lima	71	Similar to Henderson's Bush, but more prolific. Good shell bean. Recommended for short-season areas.	M, Y
Yellow Eye	bush	85-92	Plump, oval, medium-sized beans. Use dry for baking and soup.	J, Y

*To most bean growers maturity means dry seed, which can be used, if desired, for another year's crop. The figures listed here are simply designed as a guideline for reaching the most popular stage for eating, and not, in all cases, the dry bean stage.

SOURCES OF BEAN SEED

Beans are big business for many commercial seed companies. These are sources for the varieties listed.

(A) Burpee Seed Company
Warminster, Pennsylvania 18991
Clinton, Iowa 52732
Riverside, California 92502
Catalog free.

(B) DeGiorgi Company
1411 Third Street
Council Bluffs, Iowa 51501
Catalog 66¢.

(C) Farmer Seed and Nursery Company
818 Northwest 4th Street
Faribault, Minnesota 55021
Catalog free.

(D) Henry Field Seed and Nursery Company
407 Sycamore Street
Shenandoah, Iowa 51602
Catalog free.

(E) Gurney Seed and Nursery Company
Yankton, South Dakota 57079
Catalog free.

(F) Joseph Harris Company
Moreton Farm
3760 Buffalo Road
Rochester, New York 14624
Catalog free.

(G) H. G. Hastings Company
Box 42-74
Atlanta, Georgia 30302
Catalog free.

(H) Herbst Brothers Seedsmen
1000 North Main Street
Brewster, New York 10509
Catalog free.

(I) J. L. Hudson Seedsman
World Seed Service
Box 1058
Redwood City, California 94064
Catalog free.

(J) Johnny's Selected Seeds
Albion, Maine 04910
Catalog 50¢.

(K) J. W. Jung Seed Company
339 South High Street
Randolph, Wisconsin 53956
Catalog free.

(L) Kilgore Seed Company
1400 West First Street
Sanford, Florida 32771
Catalog free.

(M) D. Landreth Seed Company
2700 Wilmarco Avenue
Baltimore, Maryland 21223
Catalog $1.

(N) Earl May Seed and Nursery Company
100 North Elm Street
Shenandoah, Iowa 51603
Catalog free.

(O) Natural Development Company
Box 215
Bainbridge, Pennsylvania 17502
Catalog 25¢.

(P) Nichol's Garden Nursery
1190 North Pacific Highway
Albany, Oregon 97321
Catalog free.

(Q) L. L. Olds Seed Company
P. O. Box 7790
Madison, Wisconsin 53707
Catalog free.

(R) Park Seed Company
Greenwood, South Carolina 29647
Catalog free.

(S) Reuter Seed Company
320 North Carrollton Avenue
New Orleans, Louisiana 70119
Catalog free.

(T) R. H. Shumway Seedsman
628 Cedar Street
Rockford, Illinois 61101
Catalog free.

(U) Stokes Seeds
Box 548
Buffalo, New York 14240
Catalog free.

(V) Thompson and Morgan
Box 100
Farmingdale, New Jersey 07727
Catalog free.

(W) Otis S. Twilley Seed Company
Box 65
Trevose, Pennsylvania 19047
Catalog free.

(X) Unwins
Box 9
Farmingdale, New Jersey 07727
Catalog free.

(Y) Vermont Bean Seed Company
Box 308
Bomoseen, Vermont 05732
Catalog 25¢.

(Z) Wyatt-Quarles Seed Company
Box 2131
Raleigh, North Carolina 27602
Catalog free.

HARVESTING, DRYING,
AND THRESHING

The correct time to harvest beans depends on how you want to eat them. For instance:

- Snap beans are easy. When you think they look ready to pick, they probably are. They grow rapidly, so the picking quickly gets to be selective, and the wisdom of planting at two-week intervals is apparent. Freeze, can, or dry. (In the South, dry snap beans are called leather breeches.)

- Beans at shell stage have passed the hurry-up point, so pick them in free time and shell at leisure. Freeze or can.

- Dry beans are easiest — no freeze bags or canning jars. With pole varieties which ripen unevenly, pick the dry, crisp pods as they appear, or pull the whole plant if most of the pods are dry at once. Hang or stack loosely, off the ground, to let the roots and green leaves dry to a crisp state.

Hand-picked dry pods can be kept dry in burlap bags hung in a loft. Later on they can be crushed by hand rubbing to release the beans. In the field, mature bush beans can be further dried the old-time way — on stakes. Some bush varieties retain green leaves even when pods are all dry. Stacking whole uprooted plants is one way to hurry up leaf drying, but it is tough, stoop labor to pull the roots up. They will be dry by the time the stacking poles are ready.

I have used two types of racks to hold ripe bean plants in the field. One type consists of two slim, 6-foot poles driven into the ground about 8 or 10 inches apart. A narrow board is nailed across, connecting the poles about a foot from the ground. Sometimes baling wire is used instead of the board.

The bean plants, a good cluster in each hand, are laid on the lower crosspiece from opposite sides. They will balance one another pretty well, and will catch on the poles to remain in place. When the pile reaches the tops of the poles, wire or binder twine is wound across to tie the stack firmly.

Another type of stack uses one pole, with two pieces of board nailed in crisscross fashion a foot off the ground. In this stack system, the bunches of plants are laid in spiral fashion around the pole. Each new bunch holds the roots of the previous bunch, so the roots are

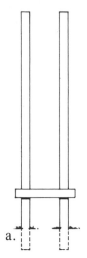
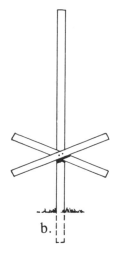

Two methods of stacking uprooted bean plants to speed leaf drying: (a) Drive two poles into the ground 8 to 10 inches apart, nail a narrow board across, and stack plants *from opposite sides so they balance and hold one another. (b) Drive one pole into the ground, nail two boards crisscross, and stack bean plants in spirals around the pole.*

against the pole. The tops are secured with wire or twine, or even locked into each other's roots with no tying needed.

At this point in the operation you may notice some nodules on the bean roots. These are the rhizobium growths, with the soil-improving nitrogen in them. Cutting off the bean plants would have left the nitrogen in the soil, but would also have made the pulling and stacking more difficult. Anyway, if the dry and threshed bushes are composted or tilled in, the nitrogen will be back where it is wanted.

When leaves are dried to a crisp, the whole stack can be carried under cover to wait for threshing. Of course, if there is plenty of space available to keep the stacks apart, drying could be done inside. We just never had all that room.

Threshing is simply done by beating a pile of the crisp, dry plants and pods to break out the beans. On a dry, tight floor or on a tarp, the pile can be beaten with a jointed flail, or a whippy branch with the heavy end doing the threshing. One caution for new threshers — do not beat the beans against the floor. Make sure there is chaff under the flail end. Break the pods, not the beans.

Actually, I like my own threshing outfit. It is a cone of bean-tight burlap, 5 feet long, 3 feet wide at the top and tapering to a 6-inch opening at the bottom. I hang it on a branch, tie the bottom, and load at the top. Two clubs, beating on opposite sides, beat the beans out of the pods. They fall to the bottom and are released into a container

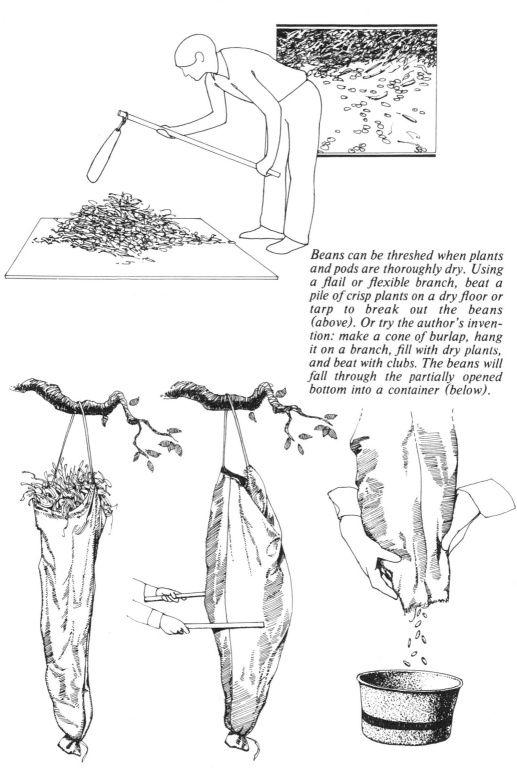

Beans can be threshed when plants and pods are thoroughly dry. Using a flail or flexible branch, beat a pile of crisp plants on a dry floor or tarp to break out the beans (above). Or try the author's invention: make a cone of burlap, hang it on a branch, fill with dry plants, and beat with clubs. The beans will fall through the partially opened bottom into a container (below).

through the partially opened bottom. This keeps most of the chaff in the bag, which is upended to empty it and start another load. This is a good system for many.

Cleaning dry, chaffy beans is made easy with wind, either from a fan or from nature whistling air around the corner of a barn or shed. Beans are heavier, and the chaff bits and dust just blow away when the beans are dropped from a container to a blanket or tarp. It may take several drops, and the wind will probably be cold, but the beans will be clean.

STORAGE

An important element in storing dry beans is eliminating moisture from the storage containers. Natural air drying of small quantities is usually sufficient, and slow drying is preferred. Some Indians used to boil their bean harvest (for food, not seed), then dry them for storage. That would be a good practice for us today. Dry and cold is better than dry and hot, but either is better than even slight moisture at any temperature. Fully dry bean seed can be sealed if kept from rapid changes of temperature.

Even glass canning jars with tight-fitting screw tops will "breathe" or exchange air when either temperature or air changes, especially if changes are rapid. For example, beans stored at 70° F. will "inhale" air when the temperature drops to, say, 50° F., and that cooler air will "condense out" moisture inside the jar. That, in itself, is not so bad, really, because beans which are stored in room air having approximately 65 percent relative humidity, and a temperature of 70° F. will dry to an equilibrium or balanced moisture content *in the seeds* of about 12 percent. That is nearly down to a perfect 10 percent moisture level for long-time storage.

One way to hold dry beans in a safe, low moisture condition is to place packets of a dessicant material in the tight containers. Silica gel and activated carbon are two materials which will absorb free moisture and hold it away from the seeds. Dry milk has been suggested for this purpose, also, but use it only in very dry air conditions. Soft plastic bags, even sealed ones, will not exclude water vapor, and are not recommended.

Store beans only when dry, keep them dry, and keep both the beans and yourself cool.

No discussion of bean storage can ignore the little bugs that come out of the little holes in the beans. These are the bean weevils: tiny, less than 1/8 inch long, hard-shelled, gray, active pests. Containers such as metal preserving jars or coffee cans with plastic tops that will exclude weevils (or contain them), but will permit "breathing" when air pressure and temperature changes occur are considered the best. Soft plastic bags will not contain the weevils, or keep out water vapor.

The summertime adult weevil lays eggs on bean pods. The larvae eat beans a while, then drill into the forming seeds. As many as a dozen may get into one seed, where they form pupae which are covered by the skin of the seed and remain for months, unseen, until they change to adults and cut their way out of the dry seed, leaving a perfectly round hole which is often the first evidence of their presence. They are capable of reducing whole jars of beans to dust, and the adults fly to any open containers of beans and repeat the egg laying-to-adult cycle several times a year.

Some evidence seems to implicate animal manure as a refuge for weevils in gardens, but most controls are applied directly to the dry beans, and not to the insects' breeding territory. The activity of the weevils can be slowed to a stop with cold. They become inactive at a temperature of around 50° F. and are killed by freezing to 0° F. for a day or so. Some writers recommend heating to a temperature of 145° F. for an hour. This is fine for food beans, but poor practice for seed beans.

My own tests with an old southern control method of putting some snuff in each jar seem to prove it a good idea. I placed one teaspoon of snuff in a small cup on top of a half-full pint jar of beans infected with active weevils. The snuff did not touch the beans, but within a few hours at room temperature, all the weevils were apparently dead, and none revived after removal of the snuff. Some weevils had climbed into the cup and died on the snuff. Beans not drilled by weevils in the jar were viable and sprouted normally.

Though this is good evidence for further testing, it falls short of being conclusive. Not until it can be shown that the pupae, as well as the adult outside the seed are killed, can it be deemed a full control. A dusting of malathion has been used in a manner similar to the snuff control, but no mention of pupae kill has been noted in this case either.

GROWING AND COOKING BEANS

COOKING BEANS

COOKING BEANS

GENERAL REMARKS

There are many different varieties of beans in just the genus *Phaseolus,* the focus of this book, and no recipe tastes exactly the same when made with two different kinds of bean — even if they're both snap beans, wax beans, or dry beans. Flavor is such an elusive characteristic that to discuss it is virtually impossible. Suffice it to say that some beans taste "beany," others "nutty," some seem more fragrant than others, and some exhibit bright colors that seem to enhance their flavor.

Bean textures also vary from "crisp" and "brittle," words often used to describe wax and snap beans, to "mealy" and "soft," for dry beans. Since there is so much variation, I can only urge you to grow (or purchase) as many different beans as you can, sampling their distinctive flavors and textures as you cook them different ways. Try them in all stages — snap, shell, and dry — to find your favorites.

DRY BEANS

Before cooking, soak dry beans overnight in cold water, allowing 2½ to 3 cups of water for each cup of beans. Bring the water to a boil, immediately reduce heat, and simmer gently until beans are tender. Older beans often require more cooking time than young ones, so watch them carefully as they cook to determine when they're tender. (The age-old trick, the best test for doneness, is to spoon out a few beans and blow on them; if their skins wrinkle up, they are done.) If cooked too long, some varieties of dry beans become mushy and unpalatable. Here also you must experiment. Be sure beans are covered with water during soaking, and that they have ample water when cooked during stove-top simmering or oven baking. The recipes that follow do not state specific amounts of water to use since this varies with type of beans, method of heating, etc.

Season initial cooking water with onions or herbs, but parboil beans before adding salt, fat, or meat. They will become tender faster by this method.

Here are some approximate times for simmering presoaked dry beans in water to cover:

Black beans — 2 hours
Cranberry beans — 2 hours
Great Northern beans — 1 to 1½ hours
Kidney beans — 2 hours
Lima beans — 1 hour (Baby Limas — 45 minutes)
Navy (Pea) beans — 1½ to 2 hours
Pinto beans — 2 hours

Dry beans swell up when soaked. Keep these equivalents in mind as you plan a meal:

1 cup dry beans = 2 to 3 cups cooked beans
1 cup dry beans = 4 servings
2 cups dry beans = 1 pound

FLATULENCE

The fact that dry beans cause gas for some people should not be news to anyone. But just what it is that causes this socially unacceptable side effect *is* news, and is the basis of various studies currently being conducted in the United States, Canada, Europe, Asia, and Africa. Researchers have thus far concluded that the sugars in beans (oligosaccharides) are one of the main causes of flatulence, but that other components must be studied as well, especially carbohydrates and bacteria or combinations of bacteria in the body itself.

Dr. Brent Skura, microbiologist and food scientist at the University of British Columbia in Vancouver, received a $36,000 grant from the Canadian Agriculture Ministry to study gas in beans. Starting in early 1980, and working with human volunteers and vitrocultures, Dr. Skura hopes to develop a simple system to study what changes occur in the body after beans are consumed, and the chemically quantitative factors that cause flatulence.

Undoubtedly, in days to come newspapers and magazines will report any conclusive findings and tell us what, if anything, we can do when cooking beans to cut down on this side effect. Meanwhile, from a number of sources come these thoughts:

1. Soaking beans overnight removes some of the oligosaccharides, and, as many people correctly maintain, if you throw away the soaking water, you will lessen the problem. Unfortunately, the soaking water also contains important vitamins and nutrients that are thus lost. In the interest of possible discomfort but more nutrition, use the soaking water to cook dry beans.

2. The longer you soak beans and the slower you cook them, the easier they are to digest. So do not be in a hurry to speed up the cooking process. (You may read suggestions for short soaking and then frying in oil, as was tried in Brazil, or the USDA's fast-cook method of boiling unsoaked beans in water for two minutes, soaking for one hour, and then cooking. Try these also, and draw your own conclusions.)

3. Adding baking soda (sodium bicarbonate) to the soaking or cooking water is widely practiced as a gas preventative. This is not recommended; soda does not work on the culprit sugars, and it destroys many valuable nutrients.

4. Some cooks claim to have reduced or solved the flatulence problem by adding castor oil or safflower oil to the bean cooking water. Others suggest dill weed or parsley. Some devotees of Mexican cooking claim that small quantities of epazote (worm-seed), an herb that grows wild in many parts of the country, reduces gassy effects. (You can obtain seeds from Horticultural Enterprises, Box 34082, Dallas, Texas 75234.)

NUTRITION

Beans provide many substances that the human body requires. Early man somehow understood, or at least practiced, the correct nutritional mix of beans and corn, or beans and wheat, which provided complete vegetable protein. Protein is important (read *Diet for a Small Planet* by Frances Moore Lappé* for more information), and the FDA confirms the fact that legumes contain more and better-quality protein than other plant sources. But a current nutritional trend suggests that we have become too hung up on protein, and that the daily amounts recommended by the National Research Council — .8 gram per kilogram of body weight (40 grams for a 110-pound person; 60 grams for someone weighing 165 pounds) — may be in excess of what the body needs. Watch for newspaper reports on this in the future.

Here is a brief summary of some of the major nutrients in beans. We have included raw and cooked beans to show you how the values change from one form to another.

*Published by Ballantine Books, a division of Random House, New York; first printing September 1971, revised edition April 1975.

Nutrients in Edible Portions Per Pound

	Calories	Grams/Protein	Grams/Carbohydrates	Milligrams/Calcium	International Units/Vitamin A	Milligrams/Iron	Milligrams/Sodium	Milligrams/Ascorbic Acid
Dry White Beans								
raw	1,542	101.2	273.1	653	*	35.4	86	**
canned w/pork and sweet sauce	680	28.1	95.7	286	**	10.4	1,724	**
canned w/out pork	544	28.6	104.3	308	270	9.1	1,533	9
Dry Red Beans								
raw	1,556	102.1	280.8	499	90	31.3	45	**
canned	408	25.9	74.4	132	Trace	8.2	14	**
Pinto, Calico, and Red Mexican, raw	1,583	103.9	288.9	612	**	29.0	45	**
Black and Brown, raw	1,538	101.2	277.6	612	140	35.8	113	**
Lima Beans								
Immature seeds, raw	558	38.1	100.2	236	1,320	12.7	9	130
Fordhook, frozen	463	28.1	88.5	104	1,040	8.6	585	101
Baby Limas, frozen	553	34.5	104.3	172	1,000	12.7	667	85
Mature seeds, dry, raw	1,565	92.5	290.3	327	Trace	35.4	18	**
Green Snap Beans								
raw	128	7.6	28.3	224	2,400	3.2	28	76
frozen, French-style	122	7.7	27.7	181	2,400	4.1	9	45
Yellow (Wax) Beans								
raw	108	6.8	24.0	224	1,000	3.2	28	80
frozen, cut	127	8.2	29.5	163	450	3.6	5	54

After table in *Composition of Foods*, USDA Handbook No. 8, pages 8-10. Revised December 1963; approved for reprinting October 1975.
* Indicates amount of constituent is none, or too small to measure.
** Denotes lack of reliable data for constituent believed to be present in measurable amount.

BEAN DIPS

BLACK BEAN DIP

1 medium onion, finely chopped
1 small clove garlic, minced
½ stalk celery, minced
2 tablespoons lard
1 cup black beans, cooked
1 teaspoon salt
⅛ teaspoon pepper
2 tablespoons Worcestershire sauce
⅛ teaspoon hot pepper sauce
1 tablespoon sugar
Salt to taste

Sauté onion, garlic, and celery in lard until wilted. Combine with beans and other ingredients and mix in blender until well pureed. Simmer over low heat for 20 to 25 minutes, being very careful it does not scorch. Add milk if puree gets too dry, or until it has the consistency of a dip. Serve hot or cold with potato chips, crackers, or tortillas. Makes 1½-2 cups.   *Roger F. Sandsted, Cornell University, Ithaca, New York*

TANGY BEAN DIP

1 cup boiled Navy beans
3 teaspoons horseradish
½ teaspoon minced onion
5 drops hot pepper sauce
½ cup chopped ripe olives
2 tablespoons mayonnaise

Drain and mash beans. Add horseradish, onion, hot pepper sauce, olives, and mayonnaise. Mix thoroughly. Makes about 1½ cups.

Michigan Bean Commission

BAKED BEAN DIP I

1 cup mashed baked beans
3 teaspoons horseradish
4 to 5 drops hot pepper sauce
½ cup chopped ripe olives
2 tablespoons mayonnaise

Combine ingredients and keep refrigerated until serving. Makes 1½ cups.

BAKED BEAN DIP II

1 cup mashed baked beans
1 teaspoon lemon juice
1 teaspoon catsup
¼ cup minced onion
⅛ teaspoon celery salt

Combine ingredients and keep refrigerated until serving. Makes 1¼ cups.

Michigan Bean Commission

HAM AND BEAN DIP

1 cup dry Baby Lima, Navy, Great Northern, or
    other white beans
2½ cups water (add more as beans cook, if
    needed, to keep covered)
1 cup ham
½ cup chopped celery
1 medium onion, chopped
½ teaspoon salt

Wash and sort beans and place in slow cooker with other ingredients, except salt. Cook on low setting 12 to 15 hours. (Or, presoak beans overnight, add to other ingredients except salt, and cook until beans are soft.) Drain and reserve liquid. Put cooked ingredients in cone-type ricer or food mill and press out pulp. Blend pulp thoroughly and add salt to taste. To get maximum ham flavor in dip, remove ham before straining, chop finely by hand, and add to pulp separately. If dip is too thick, add bean liquid for desired consistency. Serve hot or cold with whole grain chips or crackers. Makes about 1½ cups.

Color-flavor substitutions for above recipe:

Brown Bean Dip — Use Pinto, Christmas Lima, or other brown beans; 1 medium green pepper, chopped, instead of celery or onion; ½ cup pork fat, pork rind, and/or ham instead of just ham.

Red Hot Bean Dip — Use Dark Red Kidney beans or other red or pink beans; 1½ cayenne peppers, chopped, instead of celery; ½ cup pork as for Brown Dip; after sieving add ⅓ cup catsup and salt to taste.

Dark Bean Dip — Use Black Turtle or other black beans; omit celery; ½ cup pork as for Brown Dip; after sieving add 2½ tablespoons Worcestershire sauce and salt to taste. For darkest color, simmer bean liquid until it becomes thick sauce and add to pulp.

Richard D. Kahoe, Wheatfield, Indiana

GARDEN DIP

3 tablespoons oil
1 small onion, chopped
½ cup chopped green pepper
1 cup chopped eggplant
½ cup skinned, chopped, fresh tomatoes
1 cup cooked dry beans
Juice from ½ lemon
Salt and pepper

Heat oil in large frying pan. Sauté onion and green pepper. When they are just tender, stir in eggplant, tomatoes, and beans; cook slowly over low to medium heat until juice forms; cover, and cook on low heat for 15 minutes, or until all vegetables are tender and juices have blended. Add lemon juice and stir. (Drain off liquid if very soupy.) Add salt and pepper to taste. Serve with pita bread and cheese.

BEAN SOUPS

CREAM OF LIMA BEAN SOUP

Serves 6-8

Soak 1 pound of dry large lima beans (such as King of the Garden) in cold water overnight. In the morning boil beans for 10 to 15 minutes. Drain, save liquid. Place beans in cold water for a few minutes, then slip off their skins. (This sounds like a job, but they come off easily.) Discard skins and return beans to original broth. Cook until very tender, then mash in the broth with a potato masher until smooth. Add milk to desired consistency, butter, salt, and pepper to taste. Serve hot.

Carol McFadden, Nova, Ohio

BEAN CHOWDER I

Serves 12-14

¼ pound salt pork, diced
1 large onion, diced
2 pounds dry red, white, or black beans, soaked
    overnight
3 cups corn kernels
1½ teaspoons salt
1 quart milk
Black pepper to taste

Cook pork over medium heat, stirring occasionally, until it turns brown, but does not become dry. Sauté onion in pork fat until tender, but not brown. Add soaked beans and soaking water and simmer 1 hour or until tender. Add corn, salt, and milk. Adjust seasonings and simmer to ripen the flavor. Set aside, cool, and serve hot a day or two later with oyster or common crackers.

BEAN CHOWDER II

Serves 10

1½ pounds dry beans
2 teaspoons salt
1 cup diced potatoes
½ cup chopped onions
1½ teaspoons flour
1 tablespoon butter
1 cup stewed tomatoes
⅓ cup chopped green pepper
1½ cups milk

Soak beans overnight. Next day boil 30 minutes. Add salt, potatoes, and onions. Cook 30 minutes. Mix flour with butter and add to chowder. Add tomatoes and green peppers. Cook over low heat for 15 minutes longer, watching carefully to avoid sticking on bottom. Stir in milk and serve with crackers.   *Helen Handell, Saunderstown, Rhode Island*

BEAN CHOWDER III

Serves 4

1 cup dry white beans
¼ pound salt pork, cubed
1 large onion, chopped
1 large raw potato, cubed
Chopped celery tips and leaves
1 tablespoon flour
Milk (optional)
Salt and pepper to taste

Soak beans overnight, bring to boil, and simmer 2 hours. Brown pork and onion in pan and add to beans. Simmer 1½ hours longer. Add potato and celery to beans and simmer ½ hour, or until vegetables are tender. Thicken with flour and add milk if desired to replace water lost during simmering. Season with salt and pepper.

SAUSAGE BEAN CHOWDER

Serves 6-8

1 pound bulk pork sausage
1 large onion, chopped
1 pound kidney beans, cooked
1 quart water
1 quart tomatoes
1 bay leaf
1½ teaspoons salt
½ teaspoon thyme
Pepper to taste
1 cup diced raw potatoes
½ green pepper, chopped

Brown sausage in skillet. Add onion and cook until golden. Put all ingredients except potatoes and green pepper in large kettle. Simmer 1 hour. Add potatoes and pepper. Remove bay leaf. Cook until potatoes are tender.   *Mrs. Melvin Beimfohr, Farmersburg, Iowa*

PIONEER BEAN SOUP

Serves 4

1 cup dry Navy or Great Northern beans
1 ham bone
1 medium onion, chopped
½ cup diced carrots
½ cup chopped celery
3 sprigs parsley, finely chopped
1 cup peeled, diced raw potatoes

Soak beans overnight in cold water. Bring to boil and simmer until nearly done. Drain. Put beans with all other ingredients in kettle and cover with water. Simmer until vegetables are tender.

Mrs. Melvin Beimfohr, Farmersburg, Iowa

PASTA E FAGIOLI

Serves 6

1 pound dry Navy beans
3 quarts water
1 beef bone
3 tablespoons oil
2 cloves garlic, chopped
1 onion, chopped
½ green pepper, chopped
2 cups tomatoes
½ teaspoon basil
½ teaspoon oregano
Salt and pepper to taste
1 cup macaroni, cooked (or more, for very thick
    soup)
Parmesan cheese

Soak beans overnight. Drain soaking water into large kettle, and add enough water to make 3 quarts. Place beef bone in water and bring to boil. Slowly stir in beans and simmer until tender. Heat oil in skillet and sauté garlic, onion, and green pepper. Add these vegetables to beans, along with tomatoes, herbs, seasonings, and drained macaroni. If mixture is too soupy, remove some of the beans, mash, and return them to the pot, to serve as thickening agent. Serve sprinkled with Parmesan cheese. *Vermont Extension Service*

PORTUGUESE BEAN SOUP I

Serves 6-8

2 cups dry Dark Red Kidney beans
1 onion, chopped
1 tablespoon oil
1 pound linguiça, cut in ½-inch lengths
½ teaspoon cumin seed
2 cups beef stock
2 cups tomato sauce
½ bunch parsley
2 cups finely chopped kale
2 medium raw potatoes, peeled and diced
Salt and pepper

Soak beans overnight, bring to boil, and simmer ½ hour. Meanwhile sauté onion in oil in large frying pan. Add linguiça and cook for 10 minutes over medium heat, stirring occasionally. Remove from pan, drain off fat, except for 1 tablespoon. Put cooked beans, cooking liquid, linguiça, onion, and cumin in soup pot. Add stock, tomato sauce, and if necessary, water to cover. Simmer ½ hour. Sauté parsley and kale in frying pan in reserved fat. Add to pot, along with potatoes. Simmer ½ hour longer. Add salt and pepper to taste. Serve with corn bread.

PORTUGUESE BEAN SOUP II

Serves 12-14

2 cups dry red beans (dry white or limas can also
    be used)
3 ham shanks
1 medium linguiça
1 small soup bone
1 cup tomato sauce
¼ teaspoon cumin seed
1 medium onion, sliced
3 tablespoons parsley
2 tablespoons oil
2 large potatoes, peeled and diced
Salt

Soak beans overnight. Drain, reserving liquid. Boil ham shanks,
linguiça, and soup bone in kettle, in water to cover for 30 minutes.
Add beans and cook ½ hour until tender. Add tomato sauce and
cumin. Fry onion slices and parsley in oil until tender. Add to kettle
along with potatoes and cook ½ hour until done. Salt to taste. (You
can also add other vegetables such as cabbage, watercress, or carrots.)
Serve with rice or bread.   *Maui Extension Homemakers Council*

BLACK BEAN SOUP

Serves 6-8

1 pound dry Black Turtle beans
¼ pound ham or bacon rind
1 large onion, chopped
2 cloves garlic, minced
2 tablespoons butter
2 cups tomato sauce
Salt and pepper
Sherry
Parmesan cheese

Soak beans overnight. Bring to boil and simmer with ham or bacon
rind for 2 hours or until tender. Sauté onion and garlic in butter. Add
tomato sauce and stir this mixture into the beans. Add salt and
pepper to taste. Put through blender until smooth. To serve, add

1 tablespoon sherry and sprinkling of Parmesan cheese to each bowl.
(Or garnish with a slice of lemon and a slice of hard-boiled egg. Or add
¼ teaspoon lemon juice and 1 tablespoon sour cream to each
serving.)   *Ruth Sammet, San Jose, California*

BLACK BEAN SOUP EXTRAORDINAIRE

Serves 8-10

1½ pounds dry black beans
1 ham bone
2 ounces salt pork
1 clove garlic
1 bay leaf
¼ teaspoon thyme
2 medium onions
1 small sweet green pepper
1 carrot
Salt and pepper to taste
½ cup sherry
Cooked long-grain rice
1 lemon, thinly sliced
2 hard-boiled eggs, sliced

Soak beans and bone in water overnight. Bring to boil and simmer in
same water (adding more as necessary to cover) with minced salt
pork, garlic, bay leaf, and thyme for 2 hours. Chop onion, sweet
pepper, and carrot, add to beans and simmer for 1 hour longer.
Remove bone. Put mixture through blender, a small amount at a
time, until smooth. Return to kettle and reheat. Add salt, pepper, and
sherry. In heated soup bowl drop 1 generous tablespoon of rice; then
pour bean soup over rice and garnish with thin slices of lemon and
egg.   *Catharine Teall, Dundee, Michigan*

MEXICAN BLACK BEAN SOUP

Serves 4

½ pound dry black beans
2 quarts cold water
4 tablespoons oil
1 medium onion, chopped
1 clove garlic, minced
½ teaspoon crumbled chili pepper
1 tomato, peeled and chopped
¼ teaspoon oregano
Salt and pepper

Wash but do not soak beans. Place in large saucepan with water; cover and cook gently until almost tender. Heat oil in skillet and cook onion, garlic, and chili pepper until onion is tender but not browned. Stir in tomato. Combine this mixture with beans, add oregano, and salt and pepper to taste. Simmer, covered, until beans are very tender. Puree in blender or push through sieve. Return to saucepan and heat until hot.

Kay Jenness, St. Paul, Minnesota

BLACK BEAN SOUP WITH GRAPEFRUIT

Serves 6-8

1 pound dry black beans
4 cups beef stock
2 cups water
½ pound ham, cubed
1 grapefruit, peeled and cut into small pieces (or 1
    orange, or 1 cup seeded grapes)
1 green pepper, finely chopped
1 clove garlic, crushed
¼ cup dry sherry
1 teaspoon salt

Soak beans overnight, bring to boil, and simmer until tender in beef stock and water. Add all other ingredients and simmer for 30 minutes. Serve in soup bowls with side dishes of rice. If desired, garnish with finely chopped onions marinated in equal portions of vinegar and oil.

Roger F. Sandsted, Cornell University, Ithaca, New York

TURKEY BEAN SOUP

Serves 6-8

1 pound dry beans
2 turkey legs
1 gallon water
Salt and pepper to taste
1½ teaspoons thyme
Sprig of fresh rosemary
¾ cup chopped onion
¾ cup peeled, diced carrot
1 cup chopped celery
1 cup peeled, diced potato

Soak beans overnight. Put turkey legs, water, salt, pepper, and thyme in large soup kettle. Cook over low heat for several hours or overnight. Remove meat from bones, dice and set aside; discard bones. Add beans and soaking water, rosemary, onion, and carrot to kettle. Simmer 1 to 1½ hours, until beans are tender. Add celery and potato and cook until tender. Stir in meat and serve. (If you are using beans that you know take longer than 1½ hours to become tender, cook them separately and add along with celery and potato.)

Jean Rothrock, Liberty, Pennsylvania

U.S. SENATE RESTAURANT BEAN SOUP

Serves 12-14

2 pounds dry Navy beans
4 quarts water
1½ pounds smoked ham hocks
1 onion, chopped
1 tablespoon butter
Salt and pepper to taste

Soak beans overnight. Add ham hocks and water to cover. Bring to boil and simmer, covered, for 3 hours. Braise onion in butter until light brown, then add to soup; season and serve.

HUNGARIAN BEAN SOUP

Serves 6-8

1 pound dry white beans
2 quarts water
½ pound salt pork
1 bay leaf
1 parsley root
2 ribs celery, chopped
2 carrots, peeled and chopped
3 tablespoons flour
3 tablespoons butter
2 onions, minced
1 teaspoon paprika
2 cloves garlic
Salt and pepper to taste
½ teaspoon marjoram
1 teaspoon vinegar
¼ pound egg noodles, cooked
Sour cream

Soak beans overnight, bring to boil, and simmer in 2 quarts of water and soaking liquid with salt pork, bay leaf, and parsley root for 1 hour. Add celery and carrots and cook until tender. Brown flour in butter with onion, paprika, and garlic. Add to beans. Season with salt, pepper, marjoram, and vinegar. Heat to just boiling, add cooked noodles, and serve with a dollop of sour cream.

GARDEN SOUP

Serves 4-6

1 ham bone
¼ cup chopped salt pork
1 clove garlic
Several sprigs parsley
2 onions, chopped
2 carrots, peeled and diced
3 stalks celery, chopped (or handful of celery
    leaves)
1 cup shell beans*
Few leaves each of mint, marjoram, basil,
    rosemary, and thyme
4 large tomatoes, peeled
½ pound fresh spinach or other greens, chopped
1 cup pureed winter squash or pumpkin
Salt and pepper to taste
Grated cheese

Cover ham bone with water and simmer 1 hour. Remove bone and skim off any fat from stock. Cut off any remaining bits of meat from ham bone and return them to soup. Render salt pork in frying pan as ham bone is simmering, and sauté garlic, parsley, onions, carrots, and celery lightly without browning. Add to ham broth, simmer 1 hour, then add beans, herbs, tomatoes, greens, and squash. Simmer 30 minutes. Taste and adjust seasonings if necessary. Serve hot with a sprinkling of grated cheese.

*If you do not have fresh shell beans, use ½ cup dry Navy beans soaked and cooked first in unsalted water. Soldier beans are fine also. Green or wax beans can also be used.

BEAN SALADS

HOT BEAN SALAD

Serves 6

1 cup green peas
1 cup Baby Lima beans
1 cup snap beans, French cut
½ cup mayonnaise
1 medium onion, grated
1½ teaspoons prepared mustard
¼ teaspoon Worcestershire sauce
2 teaspoons salad oil
Dash hot pepper sauce
2 hard-boiled eggs, grated

Mix beans and peas, cook in small amount of boiling water or steam until tender, about 5 minutes. Drain. Add other ingredients, stir to mix, and serve hot. (Can also be refrigerated and served cold.)

Mrs. William Hall Preston, Nashville, Tennessee,
via Mrs. Deborah R. Kahoe, Georgetown, Kentucky

COLD BEANS WITH ONIONS

Serves 4

1 cup dry white beans
3 cups water
2 cloves garlic
2 tablespoons red wine vinegar
2 tablespoons olive oil
1 teaspoon salt
Coarse black pepper to taste
Dash dry oregano
3 tablespoons finely chopped green onions or
    chives

Cover beans with water and bring to boil. Boil for 2 minutes. Turn off heat, cover pan, and let stand for 1 hour. Add garlic and cook until beans are tender, adding more water if necessary. Drain, then add vinegar, olive oil, salt, pepper, oregano, and onions or chives. Stir to blend. Chill and serve.

Louis Maraviglia, Hillsborough, California

AMERICAN SWEET SOUR BEANS

Serves 8

2 quarts wax beans
1 cup water
½ cup vinegar
½ cup sugar
2 cloves
1 stick cinnamon

Cut beans into pieces and cook until tender, about 10 minutes. Drain. Mix water with vinegar, sugar, cloves, and cinnamon and boil for 5 minutes. Pour over drained beans. Chill 2 hours or longer and serve.

THREE BEAN SALAD

Serves 8-10

Make this salad at least 12 hours before serving.

1 cup dry kidney beans
1 cup dry Great Northern or Navy beans
1 cup dry Pinto or Cranberry beans
¼ cup minced onion
¼ cup minced parsley
¼ cup each minced green and red peppers
¼ cup chopped celery
1 hard-boiled egg, chopped
½ cup vinegar
½ cup oil
1 to 2 cloves garlic, crushed
Salt and pepper
Chopped fresh parsley, savory, or dill
Tomato wedges
Hard-boiled egg wedges

In separate pots, soak beans overnight, bring to boil, and cook until tender. Drain. Cool to room temperature and combine all beans. Add onion, parsley, peppers, celery, and chopped egg. Mix together oil and vinegar, garlic, salt and pepper to taste, and pour over beans. Garnish with parsley, savory, or dill, and wedges of tomato and hard-boiled egg.

MIXED BEAN VINAIGRETTE

Serves 4-6

Grow your own beans for this recipe and cook them until tender, or buy them canned and drain off liquid.

>2 cups cooked garbanzo beans
>2 cups cooked Red Kidney beans
>2 cups cooked Baby Lima beans

Combine beans in large bowl. Make **vinaigrette marinade** as follows:

>2 teaspoons sugar
>1 teaspoon salt
>½ teaspoon fresh-ground pepper
>4 tablespoons lemon juice
>8 tablespoons wine vinegar
>1 cup salad oil

Mix sugar with seasonings. Stir in lemon juice and vinegar. Add salad oil and shake. Reserve ½ cup dressing. Pour remaining dressing over the beans, cover, and chill overnight or longer.

To garnish add to the remaining ½ cup of marinade:

>½ cup finely chopped sweet pickle
>6 tablespoons finely chopped pimiento
>6 tablespoons minced parsley

Shortly before serving, drain the beans (marinade can be used again). Toss with enough of the garnish mixture to make them sparkle. Serve on relish plate or pack into unpeeled avocado or tomato halves.

California Dry Bean Advisory Board

BEAN SALAD

Serves 8-10

1 pound dry Navy beans
1 clove garlic, crushed
½ cup oil
½ cup red wine vinegar, or ¼ cup white vinegar
    and ¼ cup wine vinegar
1 tablespoon chopped anchovy fillets
1 teaspoon drained capers
½ teaspoon basil
¼ teaspoon tarragon
¼ cup green onion, thinly sliced
4 slices bacon, cooked and crumbled
¼ cup sliced almonds, toasted

Rinse beans and cover with fresh water. Boil for 2 minutes. Turn off heat and let beans soak for 1 hour, adding more water if needed. Return to heat and simmer 2 hours, or until just tender. Drain. Combine garlic, oil, vinegar, anchovies, capers, basil, and tarragon. Pour over warm beans and mix well. Chill several hours or overnight. Just before serving mix in half the onions, bacon, and almonds. Spoon beans into lettuce-lined salad bowl and sprinkle with remaining onions, bacon, and almonds as a garnish.

Michigan Bean Commission

SUMMER SALAD

Serves 4

1 pound green beans, cooked and sliced
½ cup radishes, sliced
½ cup celery, sliced
3 green onions, sliced
3 sweet pickles, diced
1 pimiento, chopped
1 cup cooked ham, diced
1 teaspoon Worcestershire sauce
½ cup oil-and-vinegar (or your favorite) dressing

Put first 7 ingredients in a large bowl and mix in Worcestershire sauce, oil and vinegar, or dressing. Refrigerate several hours or overnight.

Mrs. Melvin Beimfohr, Farmersburg, Iowa

RICE AND BEAN SALAD

Serves 6-8

2 cups cooked rice
1 pound dry beans, cooked and drained
2 dill (or sweet) pickles, finely chopped
1 onion, finely chopped
¼ cup mayonnaise
Salt and pepper to taste

Mix all ingredients and chill. If desired, make this recipe with tuna instead of rice.

GREEN BEAN AND CAULIFLOWER SALAD

Serves 4

1 head cauliflower, or 2 packages frozen
    cauliflower
1½ pounds green snap beans, or 2 packages
    frozen beans
1 teaspoon salt
2 teaspoons pepper
1 tablespoon chopped fresh mint
1 clove finely chopped garlic
½ cup oil
2 tablespoons lemon juice

Boil cauliflower florets and green snap beans in separate pots of salted water until tender but firm; drain and cool. Sprinkle each vegetable with half of salt and pepper, followed by chopped mint and finely chopped garlic. Arrange florets and beans attractively on deep platter. Mix oil and lemon juice and pour over all. (Vegetables should be coated but not swimming.) Stir salad carefully and let marinate at least ½ hour before serving.

GREEN BEAN AND LENTIL SALAD

Serves 8

2 cups green beans, chopped fine and cooked
2 cups cooked lentils, chilled (1 cup raw)
¼ cup sugar or honey
⅓ cup vinegar
¼ cup oil
½ teaspoon salt
¼ teaspoon pepper

Drain beans and lentils. Mix together in large bowl. In separate bowl combine sugar, vinegar, oil, salt, and pepper. Pour over bean mixture. Mix gently. Cover and chill 6 hours or overnight. Stir gently and drain before serving. *Vermont Extension Service*

GREEN BEAN AND POTATO SALAD

Serves 6

1 pound green beans
1 pound new potatoes
2 medium tomatoes, cut into wedges
¼ teaspoon salt
⅛ teaspoon pepper
1 tablespoon butter
5 strips bacon, sliced into ½-inch pieces
1 medium onion, chopped
¼ cup cider vinegar
¼ cup parsley, minced

Trim and snap beans. Cook until tender. Remove with slotted spoon. Add potatoes to bean water and cook until tender, about 20 minutes. Drain, cool, peel, and slice. Pile beans in the middle of a platter, arrange potatoes and tomatoes around them. Sprinkle with salt and pepper. Heat butter and cook bacon until crisp. Add onion and cook slowly until tender. Add vinegar, heat to boiling, then pour over vegetables. Sprinkle with parsley, and toss before serving.

Jane Parker Brown, Genoa, Illinois

LIMA BEAN SALAD

Serves 2-4

2 cups cooked lima beans, drained
½ cup minced ham or chopped bologna
½ cup chopped celery
1 tablespoon vinegar
Oil-and-vinegar (or your favorite) dressing, to
     taste
Salt and pepper to taste
2 hard-boiled eggs, chopped

Mix first 4 ingredients and marinate for at least 1 hour. Just before serving, pour oil-and-vinegar dressing over mixture. Sprinkle with salt and pepper, add eggs, stir gently, and serve on lettuce.

Mrs. Melvin Beimfohr, Farmersburg, Iowa

BEAN SIDE DISHES

FIRST OF THE GARDEN BEANS

When bush beans have developed enough finger-length pods for a meal, gather them shortly before eating. Wash and stem and toss whole into a steamer basket and cook for a few minutes until they are just tender. In a small pan gently heat about ½ cup of cream with a dollop of butter, and salt and pepper to taste. Transfer beans to a serving dish and pour cream and butter mixture over them.

Catharine Teall, Dundee, Michigan

SONOMA COUNTY ITALIAN-STYLE SNAP BEANS

This cooking method works well for frozen snap beans as well as fresh, to avoid the toughness that sometimes comes with parboiling.

Cut beans into thin strips, between 2 and 3 inches long and as thin as possible. (Use a sharp knife and put a Band-Aid on your thumb to cut against.) Heat several tablespoons of oil in heavy pan, sauté garlic clove in oil, and add beans. Stir to coat with oil. Add ½ cup cold water or stock, cover tightly, turn heat to low, and cook for 6 minutes. Stir again, add salt if desired, cover, and cook 6 to 8 minutes longer. Remove from heat and serve.

Mrs. Thomas E. Cordill, Sebastopol, California

BOMBAY BEANS

Serves 4-6

1 green chili pepper (about 1¼ inches long)
1 pound fresh green beans
1 teaspoon salt
2 tablespoons butter or oil
4 slices sweet onion (about ⅛ inch thick)
¼ cup grated fresh coconut (or packaged coconut,
    but use hot water to wash off sugar)

Split chili pepper in half and cook with beans and salt in a little water until tender-crisp. Drain. Meantime, melt butter (or oil) in pan and gently sauté onion slices until soft and golden. Add coconut and cook, stirring, for 2 minutes. Stir into the cooked, drained beans. Cover and let cook over medium-low heat for 2 more minutes.

Catharine Teall, Dundee, Michigan

SWISS GREEN BEANS

Serves 6

1½ pounds fresh green beans
2 tablespoons butter
2 tablespoons flour
2 cups sour cream
1 medium onion, finely chopped
¼ teaspoon salt
Dash of pepper (optional)
8 ounces Swiss cheese, grated
½ cup slivered almonds

Steam or boil beans until tender-crisp. Then drain and place in buttered casserole. Melt butter in skillet, stir in flour. Gradually add sour cream, stirring constantly. Preheat oven to 350° F. Add onion, salt, and pepper and continue stirring until sauce is thickened. Remove from heat and stir in cheese. Pour over beans in casserole and bake for 20 minutes. Sprinkle with almonds and leave in oven for a few minutes more, until almonds are lightly toasted.

Catharine Teall, Dundee, Michigan

GREEN BEAN CASSEROLE

Serves 4

1 pound fresh green beans
6 slices bacon
½ cup diced onion
¼ cup flour
1½ cups beef stock
⅛ teaspoon black pepper
½ pound sharp cheddar cheese, grated
¼ cup toasted bread crumbs

Preheat oven to 350° F. Cook beans in water until tender. Sauté bacon and onions for 10 minutes. Remove bacon, reserving fat in pan. Over low heat, mix flour with bacon fat and onions, and cook for 1 minute, stirring constantly. Add stock and continue to cook, stirring well until mixture thickens. Add black pepper, beans, cheese, and crumbled bacon slices. Mix well and put in greased 1-quart casserole. Sprinkle bread crumbs on top, dot with butter, and bake 30 minutes.

GREEN BEANS VINAIGRETTE

Serves 2

2 slices bacon
2 cups fresh green beans, cooked and drained
     (save liquid)
1 cup minced onion
1 tablespoon flour
½ cup bean liquid
¼ cup water
¼ cup vinegar
2 tablespoons sugar or honey
Salt and pepper to taste

Brown bacon until crisp; remove from pan. Sauté onion in bacon fat over low heat until golden; stir in flour. Add liquids slowly, stirring constantly. Add sugar and seasonings and bring to a boil. Stir in beans and cook until heated through. Sprinkle with crisp bacon pieces.

Catharine Teall, Dundee, Michigan

GREEN BEANS IN SOUR CREAM

Serves 4-6

1½ pounds fresh green beans
4 slices bacon
¼ cup chopped onions
2 tablespoons chopped parsley
2 tablespoons flour
1 teaspoon salt
1 tablespoon cider or wine vinegar
1 cup sour cream

Cut up beans and cook by preferred method until tender. In large skillet fry bacon until crisp. Remove bacon from skillet, drain, and crumble. In same skillet, using 2 tablespoons bacon fat, sauté onions and parsley until tender. Add flour and salt and cook, stirring, 1 to 2 minutes. Remove from heat. Toss cooked beans with vinegar. Add beans and sour cream to onions in skillet, stirring until blended. Heat thoroughly, but do not boil. Before serving sprinkle with crumbled bacon.

TOMATO STEWED BEANS

This recipe need not be made according to any special amount of vegetables. It is an excellent way to make use of end-of-the-season produce from your garden. The main ingredient is tomatoes. Wash and dry a large panful, remove stems, hard cores, and damaged spots, but do not peel. Cut into pieces and cook on low heat, stirring to prevent sticking, until juice begins to appear. When very soft put through food mill to remove seeds and skins. Return to low heat and cook gently for several hours to reduce volume. Meanwhile, search the garden for remaining string or shell beans (and/or celery, carrots, onion, cauliflower, peppers, zucchini, corn, etc.). When tomato liquid reaches desired thickness, add vegetables and cook on low heat for about 30 minutes. Add a bit of oregano or sweet basil and salt, and serve as a stewed vegetable; or fill sterilized jars, process in canner, and use as instant soup. *Alberta Wood, Belfast, Maine*

POTATO-GREEN BEAN CASSEROLE

Serves 4

1½ cups cooked fresh green beans
1 cup diced raw potatoes (German Fingerlings,
    preferably)
1 medium onion, chopped
1 tablespoon butter or fat
1 tablespoon flour
1 cup milk
1 cup diced or shredded cheddar cheese
½ teaspoon salt
¼ teaspoon pepper
¼ cup chopped fresh parsley
1 teaspoon fresh (or ½ teaspoon dried) savory
Buttered crumbs for garnish

Preheat oven to 350° F. In buttered casserole combine beans and
potatoes. Sauté onion in butter, stir in flour, and gradually add milk.
Cook until thickened. Stir in cheese and continue stirring until
melted. Add salt and pepper. Stir in parsley and savory. Pour mixture
over vegetables, sprinkle with buttered crumbs, and bake uncovered
for 30 minutes, or until brown.   *Ruth L. Putnam, Syria, Virginia*

GREEN BEANS AND JERUSALEM ARTICHOKES

Serves 2

2 cups sliced fresh green beans
½ cup Jerusalem artichokes, diced
3 tablespoons butter
½ cup mushrooms, sliced
Salt and pepper to taste

Cook beans in salted water until tender-crisp. Sauté diced artichokes
in butter for 3 minutes, add mushrooms, and sauté 2 minutes longer.
Mix all together and season.   *George F. Gens, Norwalk, Ohio*

SUCCOTASH

The North American Indians mixed beans with corn, forming this basic, nutritionally complete dish that has developed with its own variations over the years and throughout different parts of the country. Some people know it strictly as a combination of lima beans and corn; others prefer snap beans with corn; and yet others like the textures of dry beans and corn, or the colors of red beans and corn. And who's to say the mixture couldn't include a third vegetable as well? Here's a basic recipe using corn and fresh lima beans to serve 4:

> 1 cup fresh lima beans, cooked
> 1 cup fresh corn kernels, cooked
> 2 to 3 tablespoons butter
> Salt and pepper

Combine beans and corn in steamer and steam together over boiling water for several minutes. Remove to warm serving dish, add butter, and season to taste.

PLYMOUTH SUCCOTASH

Serves 16-20

This is hardly what you would call a side dish, but an old-time variation on the succotash theme.

> 1 quart dry Navy beans
> 4 to 6 pounds corned beef
> 4- to 5-pound fowl
> 1 turnip, thinly sliced
> 6 potatoes, thinly sliced
> 2 quarts hulled corn, cooked
> Salt to taste

Soak beans overnight. In large kettle boil meat and chicken together in water to cover. (You can do this the day before serving.) When tender, remove from liquid. Cook turnip and potatoes in meat-and-chicken stock until tender. Cook the soaked and drained beans in water to cover until soft enough to mash. Cut meat into 1½-inch cubes. Add meat and beans to vegetables and stock. Then add corn and water to amply cover. Let ingredients cook together 1 hour. Stir frequently to keep from sticking and add more water as necessary. Season with salt, if necessary, and serve in soup bowls.

A WORD ABOUT SHELL BEANS

Many people bear a grudge against lima beans, dating perhaps to
sullen sessions around the family dining table, or encounters with an
unpalatable, colorless vegetable all too frequently served in schools
and cafeterias. A lima fresh from your garden can and should change
your mind. Along with fresh limas, consider shell (or shelly) beans, a
term for beans that have gone past the snap stage, but are not dry.
When the pod becomes soft and pliable, and feels like fine leather to
the touch, the beans inside will be at their maximum size. But they
are good to eat as soon as they can be easily shelled from the hull and
will continue to cook as shell beans as long as they can be dented with
a thumbnail. (French Horticultural, Cranberry, and Flageolet beans
make wonderful shell beans.) Use small, half-grown seeds raw in
salads.

To cook, cover with water or stock and boil 20 minutes. Shell
beans are delicious served plain, seasoned lightly with salt, bits of
bacon, or leftover meat. Onion tops and parsley make white shell
beans very attractive — add during the last few minutes of cooking.
Diced vegetables — carrots, celery, tomatoes, peppers, okra, eggplant
— all combine well with shell beans, which are good with corn in
succotash, too.

Cook the delicate white beans (Blue Lake or Yellow Eye) with
chicken stock; the darker beans (Romano, Kentucky Wonder) are
good with a more robust pork, beef, or lamb stock.

Mrs. Thomas E. Cordill, Sebastopol, California

LIMAS NEUFCHATEL

Serves 4

2 cups fresh green lima beans
¼ cup butter, cut into small pieces
Salt and pepper to taste
1½ cups milk
2 egg yolks, beaten
¼ cup soft cheese, grated

Preheat oven to 350° F. Cook beans in water until almost tender.
Drain. Turn into 1-quart casserole. Add butter, seasonings, milk, egg
yolks, and cheese, stirring gently. Bake for about 25 minutes. Stir
twice during baking. *Catharine Teall, Dundee, Michigan*

BAKED LIMA BEANS I

Serves 6-8

Preheat oven to 350° F. Fill a 2-quart casserole with cooked lima beans. Pour maple syrup over them, add a little mustard, stir, and place strips of bacon on top. Bake until brown and moisture is absorbed, approximately 20 to 30 minutes. *Naomi H. Brown, San Angelo, Texas*

BAKED LIMA BEANS II

Serves 4

4 slices bacon
1 small onion, chopped
4 cups cooked dry lima beans and their liquid
½ cup molasses
1 tablespoon prepared mustard
½ cup chili sauce

Cook bacon and pour off all but 1 tablespoon fat. Sauté onion in fat until golden. In shallow baking dish combine beans, cooking liquid, molasses, mustard, onion, and chili sauce. Stir lightly to mix. Bake at 400° F. for 1 hour until bubbly, or until beans are thoroughly tender.

LIMA BEAN CASSEROLE

Serves 6-8

1 pound dry Baby Lima beans
1¾ cups bean liquid
5 slices bacon
1 onion, thinly sliced
½ teaspoon dry mustard
¾ cup catsup
1 cup brown sugar

Soak beans overnight. Bring to boil and simmer until soft (not mushy), about 45 minutes. Drain beans, reserving liquid. In bottom of pot put 3 slices of bacon cut in strips, drained beans, and other ingredients. Cook approximately 40 to 45 minutes. Preheat oven to 350° F. Then pour mixture into baking dish and add 1 teaspoon salt, if desired. Pull up pieces of bacon to top layer; put remaining 2 slices of bacon on top. Bake 45 minutes.

LIMA BEAN TIMBALES

Serves 6-8

2 cups dry lima beans, cooked
1 cup bread crumbs
2 tablespoons melted butter
2 tablespoons chopped pimiento
2 tablespoons minced onion
½ cup chopped peanuts
½ teaspoon salt
¼ teaspoon pepper
2 eggs, beaten
½ cup evaporated milk

Preheat oven to 350° F. Press beans through coarse sieve and add them to crumbs, butter, pimiento, onion, peanuts, salt, and pepper. Stir in eggs and milk. Place in 6 individual buttered dishes and bake about 40 minutes. Serve with diced carrots and broccoli.

DRY BEAN RISOTTO

Serves 6-8

1 onion
4 tablespoons oil
10 tablespoons butter
2 cups rice
4 cups meat or chicken broth, heated to boiling
2 cups dry beans, cooked and drained
½ cup mushrooms, sliced
½ cup Parmesan cheese, grated
½ teaspoon oregano
½ teaspoon basil
1 tablespoon parsley, finely chopped
Salt and pepper to taste

Chop onion, sauté in oil and 7 tablespoons butter.* Over low heat add rice and stir constantly in butter until all grains are well coated. Begin to add boiling broth, a small amount at a time, stirring constantly until rice absorbs liquid. This will take anywhere from 25 to 35 minutes. (Be patient and keep stirring.) Add beans, mushrooms, cheese, oregano, basil, parsley, salt and pepper, and 3 more table-spoons butter. Continue stirring until all liquid is absorbed. Garnish with parsley.

* This recipe is loaded with butter. If this turns you off, consider cooking rice in broth, sautéeing mushrooms and spices in oil, and combining with beans when cooked. Add cheese, ¼ cup hot broth, stir, and serve.

OLD-TIME MAINE BEANS AND APPLES

Serves 6-8

2 cups dry Navy beans
⅓ pound salt pork, sliced and cubed
3 cups sliced, cored apples
1⅓ teaspoons salt
⅓ cup molasses
Pinch of ginger (optional)

Wash beans and cover with water. Heat to boiling, cook about 1 hour or until tender. Drain, saving liquid. Preheat oven to 325°F. Arrange beans, part of the pork, and apples in layers in a casserole dish or bean pot. Season each layer with salt and molasses. When finished, top with slices of pork. Pour 2 cups cooking liquid over top (add pinch of ginger if desired). Bake until tender for about 1½ hours. Check several times; if beans become dry, add more cooking liquid.

CREAMY NAVY BEANS

Boil Navy beans until tender. Drain, reserving liquid for soup stock. Sauté beans in small amount of butter. When they are slightly browned add enough heavy cream to coat the beans. Serve hot.

HOT CRANBERRY BEANS

Serves 4

1½ cups dry Cranberry beans
½ cup olive oil
1 large onion or 3 green onions, chopped
Salt and pepper to taste
Chopped parsley

Soak beans overnight. Cook over very low heat until tender, about 2 hours. Sauté onion in oil until wilted. Add to beans. Season with salt and pepper, and garnish with chopped parsley.

BEAN SWAGIN

Serves 14-16

Maine lumberjacks frequently ate bean swagin and johnnycake, and the recipe remains popular in many New England communities.

½ pound salt pork
2 pounds dry beans
Salt and pepper to taste
Butter

Sauté until crisp salt pork cut either in slices or in cubes. Add washed, dry beans, and water to cover. Cook slowly until beans are soft. If necessary, add more water during cooking to prevent sticking. Add salt and pepper to taste, and a lump of butter.   *Maine Department of Agriculture*

RANCH-STYLE FRIJOLES

Serves 14-16

2 pounds dry Pinto beans
2 large onions
4 cloves garlic
1 can (4 ounces) roasted green chilies
1 can (10 to 12 ounces) taco sauce
1 can (35 ounces) tomatoes, whole, peeled
2 teaspoons salt
½ teaspoon black pepper
½ teaspoon cumin seed

Soak beans overnight. Heat to boiling and simmer until tender. Dice onions, garlic, and chilies and mix with taco sauce and tomatoes. Stir into beans and liquid with seasonings and, if necessary, add enough boiling water for simmering another 1½ hours, or until beans are tender. Will freeze well for later use.   *Mrs. Melvin Beimfohr, Farmersburg, Iowa*

PINTO BEAN POT

Serves 6-8

1 pound dry Pinto beans
2 to 4 tablespoons bacon fat (or ½ cup bacon or
      salt pork, or cooked diced ham)
1 cup coarsely chopped onion
1 large clove garlic, crushed
1 teaspoon salt
½ teaspoon coarse black pepper
½ teaspoon crushed oregano
½ teaspoon ground cumin
3 teaspoons chili powder
1 cup tomato sauce

Soak beans overnight and cook until tender. Reserve liquid. Cook onion in fat or along with bacon or salt pork. Add remaining ingredients, plus 1¾ cups reserved bean liquid. Cook 5 minutes.

North Dakota Cooperative Extension Service

BORDER BEANS

Serves 6-8

1 pound dry Pinto beans
¼ to ½ pound salt pork or smoked bacon, cut into
    pieces
3 to 4 cloves garlic, chopped
1¼ teaspoons oregano
4 hot peppers, mashed
1 tablespoon salt

Place beans in pot with water to cover and bring to boil. Reduce heat and simmer 1 hour. Continue cooking 4 hours more, uncovered. After first hour, add salt pork or bacon; after second hour, add garlic; after third, oregano and hot peppers; and after fourth, add salt and simmer 1 hour more. (Total cooking time: 5 hours.) Serve hot or cold with corn bread.   *Ruth W. Auten, Orlando, Florida*

BARBECUED BEANS

Serves 6-8

1 pound dry pink or light red kidney beans
2 to 3 cloves garlic, minced
1 large onion, chopped
2 cups tomato sauce
⅓ cup bacon fat
2 teaspoons salt
2 to 4 teaspoons chili powder
¼ teaspoon powdered cumin seed
(½ pound cooked sausage or ground beef)
(1 green pepper, chopped)

Soak beans overnight. Add other ingredients and water to cover. Simmer gently in covered pot 2 to 3 hours, until beans are tender and sauce is thick and rich. Stir occasionally, and add boiling water if necessary. When done the beans should be neither dry nor soupy. (If you use the optional meat and green pepper, add them to the pot for the last 30 minutes of cooking time.)   *California Dry Bean Advisory Board*

PINQUITO BEANS*

Serves 12-15

3 pounds dry Pinquito beans
2 small ham hocks
2 medium onions
1 clove garlic
1 tablespoon oil
3 tablespoons chili powder
3 cups tomato sauce
Salt and pepper to taste

Wash beans thoroughly and let soak overnight. Next day cook in water with ham hocks until tender. While beans and ham hocks are cooking, chop onion and garlic and sauté in skillet in oil; then add chili powder and tomato sauce, and let simmer 15 to 20 minutes. When beans are half-cooked, add sauce, salt and pepper, and continue to cook until tender. (Total cooking time is approximately 3 hours.)   *Williams Markets, Santa Maria, California*

*Pinquito beans do not become as tender with cooking as other dry beans.

JACKO'S BARBECUE PINQUITOS

Serves 10-12

1½ pounds dry Pinquito beans
1 pound bacon, diced
2 green chili peppers, diced
3 medium onions, diced
2 cloves garlic, minced
Salt and pepper to taste

Cover beans with water and soak overnight. Cook 4 hours, or until tender, adding water if necessary. Fry together bacon, peppers, onions, and garlic. Add this mixture to beans, season with salt and pepper, and cook 10 minutes.   *Nichols Garden Nursery, Albany, Oregon*

SAN FERNANDO VALLEY PINQUITOS

Serves 6-8

1 pound dry Pinquito beans, washed and soaked
    overnight
2 cups sliced apples
1½ cups tomato juice
¼ cup brown sugar
2 cloves garlic
½ pound uncooked bulk pork sausage
4 small sliced onions
2 teaspoons salt
½ teaspoon pepper
½ teaspoon (or more) chili powder

Bring beans to a boil, cover, and simmer 1 to 2 hours, or until tender.
Drain. Combine remaining ingredients, being sure to mix sausage
well with apples, etc. Bring to a boil and mix with drained beans.
Allow to simmer at least 2 hours. Serve with toasted garlic bread.

Mrs. C. L. Brown of Santa Maria,
via Nichols Garden Nursery, Albany, Oregon

SANTA MARIA PINQUITO BEANS

Serves 6-8

1 pound dry Pinquito beans
1 piece bacon, diced
¼ cup diced ham
1 small clove garlic, mashed
¾ cup tomato puree
¼ cup canned red chili (enchilada) sauce
1 tablespoon sugar
1 teaspoon dry mustard
1 teaspoon salt

Cover beans with water and soak overnight. Simmer 2 hours, or until
tender. Sauté bacon and ham, add garlic, and sauté 1 to 2 minutes,
then add tomato puree, chili sauce, sugar, mustard, and salt. Drain
off most of the liquid from beans and stir sauce into the beans. Keep
warm over very low heat or in low oven until ready to serve.

Bill Bridges, Ventura, California

BLACK BEANS AND VINEGAR

Serves 4-6

½ pound dry black beans
1 bay leaf
¼ cup olive oil
1 clove garlic, chopped or crushed
1 teaspoon salt
½ cup sweet red pepper, finely chopped
⅓ cup onion, finely chopped
¼ teaspoon oregano
¼ teaspoon ground cumin
(1 pound cubed lean pork or sausage made into
    small balls)
1 teaspoon red wine vinegar

Cook beans with bay leaf until tender. In separate pan heat olive oil with the garlic, salt, red pepper, onion, oregano, and cumin and cook for 15 minutes. Then add to beans. Fry the optional pork or sausage until brown and well done, mix with other ingredients, and simmer 1 hour. Before serving add vinegar. Serve with rice.

Roger F. Sandsted, Cornell University, Ithaca, New York

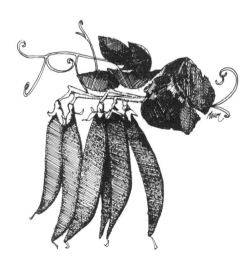

DRY BEAN CASSEROLE

Serves 6-8

1 cup dry beans, cooked (Baby Limas or mixture
    of several kinds of beans)
1 cup chopped celery
1 to 2 onions, chopped
2 cloves garlic, minced
1 green pepper, chopped
2 tablespoons olive oil
2 cups tomatoes, chopped
2 cups tomato sauce
¼ teaspoon each of thyme, marjoram, and basil
1 bay leaf
2 teaspoons soy sauce
¼ cup sherry
Grated cheese

Cook dry beans in stock. Set aside. Sauté celery, onion, garlic, and green pepper in oil until tender. Add tomatoes and sauce, cook until thick. Preheat oven to 350°F. Mix sauce, beans, herbs, and seasonings and pour into casserole dish. Add sherry and sprinkle cheese on top. Bake until bubbly and cheese has melted. Serve with bread.
        Joan Donaldson-Van Voorhees, Fennville, Michigan

MIXED BEAN CASSEROLE

Serves 6

½ cup dry kidney beans
½ cup dry Navy beans
½ cup dry black beans
1 bay leaf
¼ cup oil
1 to 2 onions, sliced
2 ribs celery, chopped
2 teaspoons celery seed
½ teaspoon dry mustard
1 teaspoon sea salt

Soak beans overnight. Next day bring to boil in water to cover with bay leaf, reduce heat and simmer gently. Heat oil in frying pan over medium heat and add onions. Sauté 3 to 4 minutes, add celery, celery seed, and mustard, and cook a few minutes more. Add vegetables to beans and continue simmering, adding water if necessary to prevent burning. When beans are tender add salt and cook another 10 minutes.

Erewhon Natural and Organic Foods, Boston, Massachusetts

LOVELY YELLOW EYES

Serves 6-8

1 pound dry Yellow Eye beans
2 teaspoons salt
¼ teaspoon pepper
2 cloves garlic, minced
1 bay leaf
2 to 4 tablespoons butter
2 onions, chopped
1 green pepper, chopped

Soak beans overnight. Add salt, pepper, garlic, and bay leaf to undrained beans and simmer gently until tender (about 45 minutes). Preheat oven to 350°F. Sauté onion and green pepper in butter and add to beans. Put into casserole and bake, covered, for 1 hour. Check from time to time and add water if necessary.

Martha W. Meyer, Lincolnville, Maine

BAKED BEANS

To avoid a display of narrow-mindedness on the subject of baked beans, we've devoted an entire chapter to this traditional mainstay of American cooking, including recipes from the South and West, as well as from the North. Experiment with different varieties of dry beans (each has its own unique flavor and texture) and different methods of sweetening or seasoning.

General Comments:

1. When cooking baked beans for a long time, be prepared to add water if they seem to be getting too dry.

2. While simmering, beans should be covered with 1 inch of water.

3. If you put salt pork on top of the beans, uncover the pot for the last hour to brown the rind.

SOUTHERN BAKED BEANS

Serves 6-8

1 pound dry beans
2 cloves garlic, minced
2 small onions, sliced
1 dried hot red pepper
1 bay leaf
3 tablespoons molasses
¼ cup catsup
1 teaspoon powdered mustard
½ teaspoon ground ginger
½ teaspoon salt
¼ cup brown sugar
½ pound smoked bacon, diced

Cover beans with water and boil for several minutes. Cover and let stand 1 hour. Add garlic, onion, pepper, and bay leaf to beans and cook together until beans are tender, 1 to 2 hours. Drain, reserving liquid. Preheat oven to 275°F. To 2½ cups of bean liquid add remaining ingredients, except brown sugar and bacon. Place beans in bean pot, stir in bacon, add liquid, and sprinkle with brown sugar. Bake for 4 hours.

BOSTON BAKED BEANS

Serves 12-14

2 pounds dry Navy or Pea beans
¼ to ½ pound salt pork, scored to rind
½ to 1 cup maple syrup
1 teaspoon dry mustard
Salt and pepper to taste
Boiling water

Soak beans overnight. Bring to a boil in the same water and simmer until tender, 1½ to 2 hours. Preheat oven to 250°F. Put beans in baking dish or bean pot, place salt pork in the center, add syrup, mustard, salt and pepper, and cover with boiling water. Bake for 8 hours.

MEETING HOUSE BAKED BEANS

Serves 6-8

1 pound dry Pinto beans
2 teaspoons salt
¼ cup finely chopped onion
1½ teaspoons dry mustard
4 tablespoons brown sugar
¼ pound bacon, cut into small pieces

Wash beans thoroughly, cover with water, soak for 2 hours, and cook over medium heat until tender, about 2 hours. Do not drain. Preheat oven to 325°F. Place beans in casserole, add salt, onion, mustard, and brown sugar. Top with bacon. Cover and bake for approximately 3 hours, adding water if necessary.

Alverda R. Swope, Leetonia, Ohio

MAINE BAKED BEANS

Serves 12-14

2 pounds dry beans
(1 medium onion)
½ pound salt pork
¼ cup sugar
½ to ⅔ cup molasses
2 teaspoons dry mustard
½ teaspoon pepper
1½ tablespoons salt
2 cups boiling water

Soak beans overnight. Bring to boil and simmer until tender, 1½ to 2 hours. Preheat oven to 300°F. If an onion is used, cut it in quarters and put on bottom of bean pot. Add beans. Cut through rind of salt pork to a depth of ½ inch, and place on top of beans. Mix sugar, molasses, mustard, pepper, and salt with boiling water. Pour this over beans and pork. If necessary, add more boiling water to cover. Bake for 6 hours. Serve with steamed brown bread or hot johnnycake.

Variations: Substitute maple syrup for molasses.
　　　　　　Add ¾ to 1 cup catsup or chili sauce and 1 tablespoon
　　　　　　　vinegar to bean mixture before baking.
　　　　　　Use ½ cup honey in place of molasses or sugar.

Maine Department of Agriculture

JOHN WITHEE'S BAKED BEANS

Serves 6-8

1 pound dry Jacob's Cattle, Yellow Eye, or
　　　Soldier beans
1 teaspoon salt
2 tablespoons dry mustard
½ teaspoon powdered ginger
4 tablespoons blackstrap molasses
⅜ to ½ pound smoked bacon

Soak beans overnight. Bring to boil in same water and simmer until tender, 1½ to 2 hours. Season beans with salt, mustard, ginger, and molasses. Put in bean pot with ¼ pound of bacon on the bottom and ⅛ to ¼ pound of bacon on top. Cover and bake at 250°F. for 8-10 hours.

GRANDMA'S HOME-BAKED BEANS

Serves 10

1½ pounds dry Yellow Eye beans
1 small onion
1 peeled apple
¼ pound salt pork
1 cup maple syrup
1 teaspoon dry mustard
1 tablespoon salt
Boiling water

Soak beans overnight. Bring to boil in same water, adding more to cover as necessary, and simmer until tender, 1½ to 2 hours. Preheat oven to 275°F. Put onion and apple on bottom of bean pot, fill half full with beans, add salt pork, and pour in remaining beans. Pour maple syrup over beans and sprinkle with mustard and salt. Cover with boiling water. Bake for 5 hours.

JOHN BROWN'S BAKED BEANS

Serves 6-8

1 pound dry Navy beans
¼ pound salt pork, cut into chunks
¼ cup brown sugar
1 teaspoon dry mustard
2 teaspoons chopped onion
½ cup dark molasses
1½ teaspoons salt

Soak beans overnight. Bring to boil and simmer until tender, 1½ to 2 hours. Drain and reserve liquid. Preheat oven to 275°F. Place beans and salt pork in casserole or bean pot. Combine cooking liquid, brown sugar, mustard, onion, molasses, and salt, and pour over beans. Cover and bake for 3 to 4 hours.

SAVORY BAKED BEANS

Serves 12-14

2 pounds dry Soldier, Great Northern, or Navy
   beans
½ pound salt pork
2 onions, chopped
¼ cup parsley, chopped
½ teaspoon dried thyme (or 1 teaspoon fresh)
2 teaspoons powdered mustard
1 scant teaspoon ginger
1 teaspoon salt
Pepper to taste
¼ cup molasses
¼ cup brown sugar
Boiling water

Soak beans overnight. Bring to boil in soaking water and simmer
gently until tender, 1 to 2 hours. Drain, reserving liquid. Preheat
oven to 275°F. Cut salt pork in half and score both halves with cuts
½-inch deep. Place 1 piece on bottom of bean pot. Cover with layer of
beans. Combine onions, parsley, and thyme. Spoon some of mixture
over beans, add another layer of beans, and repeat, layering until
both are used up, ending with beans on top. Place remaining salt pork
on top of beans. To reserved cooking liquid add mustard, ginger, salt,
pepper, molasses, and brown sugar. Pour over beans. Add enough
boiling water to cover beans. Cook, covered, for 6 to 8 hours.
Uncover for last hour of cooking.

SPICY BAKED BEANS

Serves 6-8

1 pound dry beans
½ pound lean smoked bacon
2 cups water
¼ cup celery leaves
1 cup dark molasses (or dark brown sugar)
1 onion
2 cloves garlic
2 sprigs fresh rosemary
1 teaspoon mustard seed
¼ teaspoon nutmeg
¼ teaspoon cinnamon
Salt and pepper to taste

Rinse beans and place in enameled steel pot without soaking. Cut up bacon and mix with beans. In a blender combine 2 cups water with celery leaves, molasses, onion, garlic, rosemary, mustard seed, nutmeg, cinnamon, salt and pepper. Blend for several minutes. Pour over beans and mix thoroughly. Add water so beans are covered with liquid. Place pot in refrigerator for 24 hours. Remove and bake in 250°F. oven for 8-10 hours, making more sauce and adding more water as necessary.

VEGETARIAN BAKED BEANS

Serves 12-14

2 pounds dry beans
½ cup vegetable oil
1 onion, chopped
Salt and pepper to taste
½ cup molasses
⅓ cup brown sugar
Boiling water

Soak beans overnight, bring to boil and simmer until tender, 1 to 2 hours. Place oil in bottom of casserole or bean pot, along with onion. Mix beans, salt and pepper, molasses, and brown sugar, pour into pot, and cover with boiling water. Bake at least 6 hours at 300°F.

Maine Department of Agriculture

SWEDISH BAKED BEANS

Serves 6-8

1 pound dry beans
6 cups water
1½ teaspoons salt
1 ham bone, pork chop, or small piece of salt pork
3 tablespoons brown sugar
3 tablespoons honey or dark corn syrup
1 to 2 tablespoons cider vinegar (to taste)

Soak beans overnight. Add water, salt, and ham or pork chop and cook over medium heat until tender, 2 to 3 hours. (If salt pork is used, omit salt until beans are tender; then season to taste.) When beans are cooked add brown sugar and honey or corn syrup, and vinegar. Cook for an additional 5 to 10 minutes or until beans are slightly thickened. For an extra-special dish stir in a couple tablespoons of heavy cream. Serve with boiled potatoes and crisp green salad.

Betsy Erickson Lorentzon, Tustin, Michigan

BEANS AND BEER

Serves 6-8

1 pound dry kidney beans
1 pound salt pork or 2 ham hocks
1 clove garlic, mashed
2 tablespoons oil
2 ripe tomatoes, chopped
1 green pepper, chopped
1 onion, chopped
2 stalks celery, chopped
Salt and pepper
1 can beer

Soak beans overnight. Add salt pork or ham hocks and garlic and simmer until beans are tender, about 2 hours. (Make sure beans are covered with 1 inch of water at all times.) When tender, drain and reserve liquid. Heat oil in a large skillet and sauté tomato, green pepper, onion, and celery until soft, stirring frequently. Add beans, beer, and salt and pepper to taste. Simmer 20 minutes, adding reserved bean liquid if necessary. Serve with green salad and corn bread.

Donald C. Henry, Maybrook, New York

BAKED BEANS WITH DUMPLINGS

Make baked beans according to your favorite recipe. When beans are cooked make drop dumplings, spoon on top of the beans, close the pot, and steam for 10 to 12 minutes over low heat. Do not remove the lid until time is up.

EASY DUMPLINGS

1 cup flour
1½ teaspoons baking powder
¼ teaspoon salt
½ cup milk
1 tablespoon oil

Mix dry ingredients; add liquids. Drop batter by spoonfuls onto simmering beans.

Pauline W. Schlichtling, Panama City, Florida

LEFTOVER BAKED BEANS

Here are just a few of the many ways to use baked beans a day or two after you first serve them. Since they don't keep well at room temperature, be sure to refrigerate them promptly.

BEAN SAUSAGE WANIGAN

Remove visible pieces of pork from cold baked beans and mash beans. Blend equal amounts of beans and leftover, ground-up corn bread. Add a bit of sausage seasoning to taste, or mix 1 tablespoon salt, 3 tablespoons powdered sage, 1 tablespoon summer savory, and 1 teaspoon pepper, and use some of this. Fry as sausage cakes.

BEAN PATTIES

Mash 2 cups baked beans and mix with 1 beaten egg. Shape into patties and roll in bread crumbs. Brown on both sides in hot fat or butter.

BROILED BEAN SANDWICHES

Toast 1 side of a slice of bread under broiler. Spread baked beans on untoasted side; top with sliced tomato and bacon. Return to broiler for a few minutes until bacon is crisp and beans are heated through. Or, top beans with a slice of cheese, instead of bacon, and place the tomato slice on top. Broil until cheese melts.

Delaware Extension Service

BEAN-TOMATO BAKE

Put 3 cups baked beans in baking dish. Pour 1½ cups canned, cooked, or stewed tomatoes over them and bake at 350°F. for 30 minutes.

BEANS IN PEPPER CASES

Remove stem ends and seeds from 4 green peppers. Boil 5 minutes in salted water, then drain. Fill peppers with 3 cups baked beans and pour a little catsup over them. Place in ½ inch of hot water in baking dish. Bake at 350°F. until peppers are tender and beans are heated through, about 45 minutes.   *Vermont Extension Service*

BAKED BEANS FOR A CROWD

Cook beans in quantity to serve at family reunions, church suppers, or if you have to feed the whole town!

BAKED BEANS FOR FIFTY

6 pounds dry beans
1 pound salt pork
2 medium onions, chopped
1 cup sugar
2 cups molasses
5 tablespoons salt
2 tablespoons dry mustard
1½ teaspoons black pepper
1 pint boiling water

Soak beans overnight. In the morning parboil in soaking water until skins crack when blown upon. Preheat oven to 275°F. Score the salt pork rind, then cut up a small portion of the salt pork and place in the bottom of a large baking dish with the onions. Add the beans, bury the remainder of salt pork in the top layer of beans. Mix other ingredients with 1 pint of boiling water and pour mixture over the beans. Add enough boiling water to cover. Bake 12 hours, adding more boiling water as necessary.   *Maine Department of Agriculture*

BAKED BEANS FOR 1500 TO 2000

200 pounds dry beans
35 pounds bacon
22 #10 cans catsup
16 pounds dark molasses
50 pounds dark brown sugar
1 gallon mustard
3 cups salt

Soak beans overnight. Obtain 2 steam cookers with a combined capacity of approximately 100 to 110 gallons. (An 80-gallon and 30-gallon cooker work well.) Fry bacon and add with bacon fat and beans to cookers. Add water to cover and cook slowly until beans are soft and tender. Add other ingredients. Simmer at slow boil until done. Allow 7 hours total cooking time. *Paul L. Neff, Manheim, Pennsylvania*

BAKED BEANS FOR 2000 to 3000

This recipe was put together by Mike Mitchell of the Valley Feed and Seed Company in Brush, Colorado, where beans are baked outdoors over a fire. Mike suggests using the largest kettle you can find, and stirring the beans with a wooden boat oar to keep them from sticking.

180 pounds dry Pinto beans
60 pounds bacon
60 pounds onions
18 cups brown sugar
18 cups light molasses
36 tablespoons dry mustard
10 large cans tomato paste
150 gallons water

Soak beans in the 150 gallons of water overnight. Cook bacon separately. Add all ingredients and cook over low fire all day, stirring occasionally, and adding additional water if needed.

BEAN-HOLE BEANS

In Oxford, Maine, a Bean-Hole Bean Festival is held each year, usually the last weekend in July, and some 3000 people are fed from one of the country's largest bean holes. Here are two versions of this once-popular method of cooking baked beans outdoors.

BEAN-HOLE BEANS I

Serves 50-60

Put 8 pounds of dry beans in a 4-gallon iron pot, cover with water, hang over a fire, and boil 5 minutes, stirring constantly. Remove from fire, add 3 pounds salt pork, 1 cup molasses, a handful of mustard, and a handful of salt. Cover with a closely fitting sheet-iron cover. In dry ground, preferably gravel, dig a hole 2 feet deep and 2 feet across, and in it build a brisk hardwood fire. When the fire has burned to coals, take part of the coals out with a shovel, place the iron pot in the hole, and cover with the coals previously shoveled out. On top of the coals place the gravel originally taken from the hole. Leave the pot in the hole all day.

BEAN-HOLE BEANS II

Get an iron kettle with a cover that fits tightly down over the top and a strong bail. Dig a hole in the ground at least 3 feet deep and 2 feet wide, or larger if your kettle takes more space. Place a flat rock at the bottom of the hole; add any old scrap iron or good-sized rocks. (Lacking these, the woodsmen used old logging chains.) Fill the hole with hardwood and keep the fire going for several hours to start the burning-out process. After good coals are formed, the hole is ready for cooking the beans.

Soak or parboil dry beans as for oven baking. Place in kettle with salt pork, molasses, salt, and plenty of boiling water. Cover the kettle and attach a wire, 1½ to 2 feet long, to the kettle bail. Remove all scrap iron, rocks, or chains from the hole. Lower the kettle of boiling beans into the hole. Then surround the kettle with the coals and rocks or scrap. Cover the hole in with earth. Pack tightly and tread on it to seal. (One tiny opening may allow air to get in and result in burned beans.) The following day dig off the earth, pull out the kettle by the wire, and serve cooked beans. *Maine Department of Agriculture*

MAIN DISHES WITH BEANS

CASSOULET

Serves 12-14

Cassoulet, the baked beans of France, is a heavy dish, usually made by combining beans with poultry, sausage, lamb, or other meat. Any dry white beans can be used; French chef Julia Child prefers Great Northern beans.

2 pounds dry white beans
1 carrot, peeled and chopped
2 cloves garlic, finely chopped
¼ cup chopped celery
1 bay leaf
1 teaspoon thyme
6 onions, finely chopped
3 cloves garlic, minced
6 to 8 cups cubed, roasted meat (pork, ham,
    lamb, duck, turkey, sausage, or a
    combination of these)
3 tablespoons tomato paste
1 cup bread crumbs
⅛ pound smoked bacon, thinly sliced

Soak beans overnight. Add carrot, garlic, celery, bay leaf, and thyme and simmer in water to cover about ½ hour until barely tender, not soft. Drain, reserving liquid. Preheat oven to 325° F. Sauté onion and garlic in butter or oil until barely tender. Arrange a layer of beans on the bottom of a 4-quart baking dish. Add a layer of meat, then a layer of onions, and a layer of beans, repeating in several layers, finishing with beans on top. Mix tomato paste with reserved bean liquid and pour over mixture, until top layer of beans is barely covered. (Add boiling water if necessary for enough liquid.) Top with crumbs and smoked bacon and bake, uncovered, for 1 hour, or until all liquid is absorbed.

MEXICAN CHILI

Serves 8

1 pound dry Pinto beans
1 onion, chopped
2 tablespoons oil
2 pounds ground beef
2 tablespoons flour
1 16-ounce can tomatoes
2 cloves garlic, crushed
1 teaspoon oregano
¼ teaspoon ground cumin
1 tablespoon salt
2 tablespoons chili powder

Soak beans overnight and cook in covered pot 1½ hours. Drain. In large pot brown onion in oil. Add meat and cook until light brown. Stir in flour. Add tomatoes, beans, and seasonings. Add enough water to make a thick broth. Simmer for about ½ hour.

Mrs. Melvin Beimfohr, Farmersburg, Iowa

MILD CHILI CON CARNE

Serves 4-6

1 pound ground beef
3 tablespoons butter or fat
1 cup chopped onions
3 tablespoons flour
2 cups tomatoes or tomato juice
1 tablespoon sugar
1 teaspoon chili powder
1½ teaspoons salt
4½ cups cooked Pinto beans

Fry beef in small amount of oil until redness disappears. Remove from pan. Melt butter or fat, sauté onions until yellow, then slowly stir in flour and add additional water or stock to make a thin gravy. Put into kettle with other ingredients, cover, and simmer for an hour; add more water if necessary.

Mrs. Melvin Beimfohr, Farmersburg, Iowa

CHILI BEAN PIE

Serves 6-8 (8-10 with hamburger)

1 cup water
1 cup yellow cornmeal
3 cups chicken broth
2 tablespoons oil
1 onion, chopped
1 clove garlic, chopped
1 green pepper, chopped
(1 pound hamburger)
2 cups tomatoes, peeled and chopped
1 cup corn
2 cups kidney beans, cooked
1 tablespoon chili powder
Salt and pepper to taste
1 cup grated cheese

Bring water to boil over direct heat in top of double boiler. Mix cornmeal with chicken broth and add to boiling water, stirring until mixture boils. Cook covered in top of double boiler over boiling water for ½ hour. Cool slightly. While mixture is cooking, sauté onion, garlic, and green pepper in oil until wilted; add meat, if desired, and cook until redness of meat disappears. Preheat oven to 350°F. Add tomatoes, corn, kidney beans, chili powder, salt and pepper. Spread thickened cornmeal over bottom and sides of shallow, buttered baking dish, much the same way you would line a pan with pie crust. Add bean filling, sprinkle with grated cheese, and bake for 30 minutes.

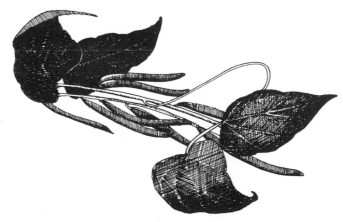

MEAT PIES

Yields 6 pies

Pastry:
¾ cup shortening
3 cups flour
1 teaspoon salt
¾ cup water (approximately)

Cut shortening into flour and salt; mix until coarse as cornmeal. Add water gradually while stirring, and shape dough into a ball. Chill.

Filling:
½ pound highly seasoned sausage meat
1½ pounds ground beef
2 onions, minced
1½ cups grated raw carrot
2 teaspoons salt
¼ teaspoon black pepper
3 cups soft cooked Pinto beans, well drained
6 teaspoons butter

Preheat oven to 375°F. Mix together all ingredients except butter. Divide pastry dough into 6 portions. Roll into circles. Put ⅙ of meat and bean mixture in each circle. Top with 1 teaspoon butter. Fold in half and seal edges. Bake on lightly greased cookie sheet about 1 hour and 15 minutes. *Mrs. Marjorie Watt, Old Monroe, Missouri*

QUICK BEAN PIZZA

Serves 4-6

¼ cup chopped onion
2 cups cooked dry beans
½ cup spicy tomato sauce or barbecue sauce
½ teaspoon hot pepper sauce
1 standard recipe biscuit dough
1 cup (4 ounces) shredded cheddar cheese

Preheat oven to 400°F. In skillet sauté onion in small amount of oil; add beans and sauces and heat to bubbling. Spread biscuit dough evenly on bottom of ungreased 9x13-inch pan. Spread hot bean mixture over dough and sprinkle with cheese. Bake 15 to 20 minutes or until biscuit dough is golden brown. *Anne Christensen, Point Arena, California*

BEAN STUFFED PANCAKES

Cook any kind of shell beans until soft. Season with butter, salt, and pepper. Mash beans. Make thin crepes or pancakes according to your favorite recipe, spread with mashed beans, add chopped lettuce and onions, roll and serve. This makes a one-dish meal that children enjoy.

June Bester, Butler, Pennsylvania

CHEESE AND KIDNEY BEAN LOAF

Serves 6

3 cups cooked kidney beans
1 cup (½ pound) grated cheddar cheese
1 small onion, minced
1 teaspoon salt
⅛ teaspoon pepper
1 egg
¼ cup milk
Tomato sauce

Preheat oven to 350°F. Mash beans or run through food mill or sieve. Combine with other ingredients except tomato sauce and put in greased 8x5-inch loaf pan. Bake 30 to 40 minutes. Remove from oven and let sit a few minutes. Remove from pan and serve with heated tomato sauce. (If some of the loaf is left over, the following day slice thinly and cook in butter or oil, a few minutes on each side, until crispy. It also makes good sandwiches.) *Vermont Extension Service*

LIMA BEAN AND SAUSAGE CASSEROLE

Serves 6-8

1 cup dry lima beans
2½ cups water
1 teaspoon salt
3 medium carrots, sliced
½ pound sausage
2 tablespoons chopped onion

Boil beans gently in salted water for 2 minutes. Cover and let stand 1 hour, then boil in same water for 30 more minutes, adding carrots during last 5 minutes. Do not drain. Preheat oven to 350°F. If bulk sausage is used, blend onion with it, shape into patties, and brown in frying pan. If link sausage is used, cut in ½-inch slices and brown with the onion. Add the cooked sausage and 2 tablespoons drippings to beans. Bake in covered casserole for about 1½ hours. If necessary, add extra boiling water during baking.   *Vermont Extension Service*

LIMA BEAN AND PEANUT ROAST

Serves 6-8

2 cups mashed potatoes
1½ pounds dry lima beans, soaked overnight and
    cooked until soft
1 cup roasted shelled peanuts, chopped fine
½ cup milk
1 beaten egg
1 teaspoon chopped onion
1 teaspoon salt
¼ teaspoon paprika
⅛ teaspoon white pepper

Preheat oven to 375°F. In buttered casserole put a layer of potatoes, a layer of beans, and then a layer of peanuts. Repeat until all are used. Mix milk, egg, onion, and seasonings and pour over mixture in casserole. Bake for 30 minutes. Serve with tomato sauce or cheese sauce.   *Catharine Teall, Dundee, Michigan*

HOPPING JOHN

Serves 4

This recipe is an old favorite in the South, but has made its way into homes in the North and West as well.

½ cup dry Pinto beans or black-eyed peas
2½ cups ham broth
½ cup chopped, cooked ham
½ cup rice, uncooked (*not* Minute Rice)
Salt and pepper to taste

Soak beans in water overnight and drain. Pour in ham broth and cook gently until tender. Add ham and rice and cook, covered, until rice is done (20 to 30 minutes). The liquid should be almost cooked away, the rice tender. Remove cover for last 10 minutes if too much liquid remains. Season and serve.   *Vermont Extension Service*

PORK AND VEGETABLE STEW

Serves 4

½ cup dry kidney beans
½ cup dry Navy beans
6 cups water
½ pound pork shoulder
2 onions
1½ cups chopped celery
3 carrots, diced
3 small potatoes, diced
Salt and pepper to taste

Soak beans in water overnight. Bring to boil and simmer with pork for 1 hour. Remove bone, cut meat from it and dice. Return bone and meat to pot. Cook for 30 minutes. Add vegetables and seasonings and cook for 30 minutes more.

PORK WITH BEANS AND CARROTS

Serves 4-6

1 pound lean pork
2 tablespoons melted butter
½ cup onions, thinly sliced
1½ cups hot chicken broth
2 cups green beans, cut into pieces
2 cups carrots, thinly sliced
2 cups celery, diced
¾ teaspoon salt
⅓ cup cold water
2 tablespoons cornstarch or flour
1 tablespoon catsup
½ teaspoon black pepper

Cut pork into thin strips about 3 inches long. Place butter, pork, and onions in heavy kettle or large skillet. Cook slowly until lightly browned, stirring once or twice. Add chicken broth, beans, carrots, celery, and salt. Cover and bring to boil. Reduce heat and simmer 20 minutes. Mix cold water, cornstarch, catsup, and pepper into a smooth paste. Add to mixture in pan, stirring constantly, and cook 3 minutes. Serve with rice and chili sauce.

BEAN BURGERS WITH SAUCE

Serves 4-6

Soak 1 cup dry beans (Pinto, black, or Cranberry) overnight in 3 cups
water. Drain, reserving water. Grind beans in food grinder using fine
blade. Combine with the following ingredients:

  2 tablespoons whole wheat flour
  2 beaten eggs
  ¼ cup sautéed onion
  1 clove garlic, mashed
  1 cup cooked, chopped greens (kale, spinach, or
      chard)
  1 tablespoon brewer's yeast
  1 teaspoon salt
  ½ teaspoon summer savory
  ¼ cup grated Parmesan cheese

Shape into patties and cook, covered, over medium heat in well-oiled
skillet for 10 minutes. Turn, cook an additional 10 minutes, covered.
Reduce heat, add sauce (see below), and simmer about 5 minutes
longer. Serve as is or on toasted buns.

  Sauce:
  2 cups canned tomatoes
  ½ cup onions
  1 tablespoon oil
  ½ cup green pepper, chopped
  ½ cup mushrooms, chopped
  1 tablespoon sugar
  ½ teaspoon salt
  1 teaspoon basil

Drain tomatoes. Sauté onions in oil. Add onions, chopped pepper,
and mushrooms to juice from tomatoes and bring to boil in saucepan.
Let simmer until reduced by about half. Cut tomatoes into small
pieces and add with seasonings to juice. Let simmer briefly. Set aside
and add to burgers.

Anne Christensen, Point Arena, California

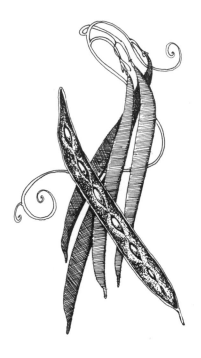

HAMBURGER WITH BEANS

Serves 4-6

1 onion, chopped
1 green pepper, chopped
1 tablespoon oil
1 pound lean ground beef
2 cups tomato sauce
2 tablespoons prepared mustard
1 teaspoon Worcestershire sauce
1 teaspoon salt
3 cups cooked dry beans and liquid

In heavy skillet sauté onion and green pepper in oil. Add meat and cook until redness disappears. Add tomato sauce, mustard, Worcestershire sauce, and salt. Mix ingredients and add the beans and liquid. Simmer for 20 minutes. For a hotter dish, add a bit of hot chili pepper.

BEEF AND BEANS

Serves 6-8

½ pound dry lima beans
½ pound dry Pinto or kidney beans
6 cups water
3 slices bacon
4 cross-cuts beef shank
2 tablespoons flour
2 large onions, chopped
1 large can Italian tomatoes
2 teaspoons salt
½ teaspoon pepper
¼ teaspoon crushed red pepper
1 teaspoon thyme
1 bay leaf

Soak beans overnight in water. Next day bring beans and liquid to a boil. Cover and let simmer for 30 minutes. Drain beans, reserving liquid. Sauté bacon until crisp in large skillet. Remove bacon, set aside. Coat beef with flour, brown on all sides in bacon drippings over medium heat. Put in 3-quart baking dish. Add beans to baking dish. Preheat oven to 325°F. Add onion to skillet and cook until soft. Stir in remaining ingredients, crushing tomatoes. Bring to boil. Pour over beef and bean mixture. Cover. Bake for 1½ to 2 hours, stirring once or twice. Add some of reserved bean liquid if mixture seems dry. Crumble bacon on top and serve.   *Jane Parker Brown, Genoa, Illinois*

PINTO BEAN CASSEROLE WITH CRUST

Serves 6-8

1 pound ground beef
¼ cup minced onion
1 egg
1 teaspoon salt
½ teaspoon pepper
Flour
2 tablespoons oil
2 cups cooked Pinto beans with liquid
2 cups cooked green beans, drained
1 cup tomato sauce
1 cup thinly sliced onions
½ teaspoon salt
Egg pastry topping

Mix ground beef, minced onion, egg, salt, and pepper. Shape into 12 balls. Roll in flour, brown in hot oil. Remove from pan. Add to pan beans, tomato sauce, sliced onions, and salt. Mix well and bring to boil. Pour hot into 2-quart casserole. Put meat balls on top. Preheat oven to 350°F. Prepare egg pastry crust (see below), and lay over top. Flute edge. Make small slits around meat balls and bake 30 to 35 minutes, until crust is lightly browned.

Egg Pastry Topping:
1 cup flour
½ teaspoon salt
¼ cup oil
1 egg
2 tablespoons cold milk

Sift flour and salt. Beat together oil, egg, and milk until thick and creamy. Pour over entire surface of flour immediately. Mix with fork to form ball and roll out.

Mrs. Carr Stolworthy, Cortez, Colorado.
From A Collection of Pinto Bean Recipes,
published by the Cortez Chamber of Commerce.

FRIJOLES RIO GRANDE

Serves 10-12

2 cups dry Navy beans, soaked overnight
1½ cups dry split peas, soaked overnight
½ pound ham hock or salt pork, chopped
6 cups water
1 teaspoon oregano
2 cloves garlic, chopped
1 large onion, chopped
¼ teaspoon ginger
1 tablespoon salt
1 teaspoon pepper
1 large onion, diced
4 stalks celery with tops, thinly sliced
4 medium carrots, thinly sliced
2 medium green peppers, sliced
2 fresh or dry hot chili peppers, chopped
1 teaspoon dry marjoram
2 cups red Burgundy wine

Cook beans and split peas until tender, about ½ hour. Add meat, water, oregano, garlic, onion, ginger, salt, and pepper and simmer 1½ hours. Add other ingredients except wine and cook 25 minutes, then add wine and cook 5 minutes more. Serve hot in stew bowls.

Michigan Bean Commission

SMOKED PORK SHOULDER AND BEANS IN CROCK POT

4 to 6 pounds smoked pork shoulder butt (picnic cut)
1 pound carrots, thickly sliced
6 small onions, quartered
1 pound Soldier beans (or other dry beans), soaked overnight in 5 cups water

Cut off outer skin of pork shoulder. Put vegetables in bottom of large pot, then add beans with soaking water. Put pork on top. Cook on high for 6 to 7 hours, or on low for 12 to 14 hours. For small crock pot, halve recipe and use a small, boned picnic butt.

Martha W. Meyer, Lincolnville, Maine

BEAN PICKLES AND RELISHES

STRING BEAN PICKLES

1 peck fresh green or wax beans
    (about 7½ pounds)
6 cups sugar
1½ cups flour
Scant ½ cup dry mustard
1 tablespoon turmeric
1 tablespoon celery seed
2 quarts vinegar

Prepare the beans as for the table. Cook slowly, with a bit of salt, until *just* tender. Do not overcook the beans because the hot pickling sauce will cook them more. Drain the beans thoroughly. Mix together the remaining ingredients and cook over *low heat* until thick. Stir almost constantly to prevent sticking. Pour the sauce over the cooked beans, fill sterilized jars with the mixture, cap, and process while hot. Process 20 minutes in boiling water to cover jars. Cool in water, remove, and store. Makes about 8 quarts.

SWEET-SOUR WAX BEANS

2 pounds fresh wax beans, cut diagonally into
    1-inch pieces
1 cup white vinegar
½ cup sugar
1 teaspoon celery seed
Pinch of ginger
1 teaspoon dried summer savory (or 1 tablespoon
    chopped fresh)
Small bay leaves

Cover the beans with water, add salt to taste and cook until just barely tender. Drain, saving the liquid; add to it the vinegar, sugar, and spices. Add more water if needed so that there will be enough liquid to fill the packed jars. Bring the liquid to boiling, add the beans, bring to a boil again, and pack the jars. Add a small bay leaf to each jar. Process. Makes 4 pints.

DILLY GREEN BEANS

2 pounds whole fresh green beans
1 teaspoon cayenne pepper
4 cloves garlic
4 heads dill
2½ cups water
2½ cups white vinegar
¼ cup salt

Pack the beans into hot sterilized pint jars lengthwise, leaving ¼ inch headspace. To each jar add ¼ teaspoon cayenne, 1 blanched garlic clove, and 1 head dill. Combine the water, salt, and vinegar; heat to boiling and pour while boiling over the beans, leaving ½ inch headspace. Cap and process. Makes 4 pints.

DILL CROCK

Fill a large crock (leaving room for vegetables) with brine made of 10 measures of water to ¾ measure of salt, adding if you wish a small dollop of vinegar (no more than ¼ measure for each 10 measures of water). Add 1 or 2 cloves of garlic and have a good supply of fresh dill on hand. Begin packing in fresh vegetables with generous layers of dill in between. Try any of the firm vegetables from your garden. Strange and marvelous things happen to green string beans if parboiled for 2 to 3 minutes and then put into a brine-filled crock. Wax beans are excellent too, along with small raw onions, raw baby carrots, whole tender pods of peas, raw cauliflower florets, cucumber chunks, etc. Cover and let stand 1 to 2 weeks. When vegetables taste pickled, remove and seal in sterilized jars in hot water bath canner.

DILLY BEANS

Purple pod beans hold their shape and work well in this recipe. Use any quantity of beans and adjust brine requirements accordingly.

Snap the ends off fresh beans and cover beans with boiling water. Cook 5 minutes, no longer. Drain and place in crock. Cover with brine made from 1 cup pickling salt, 10 cups soft water, and ½ cup vinegar. Keep beans below surface of brine with a plate. Add several pieces of dill and a few wild grape leaves. Set aside to season. Every day or so, as scum forms on the top, skim it off. When beans are nicely pickled drain, rinse, and pack into hot canning jars. Add a clove of garlic, and a piece of hot pepper if you like, and cover with freshly made brine — this time ¾ cup salt, 1 cup vinegar, and 4 quarts water. Process 10 minutes in boiling water bath.

Helen Handell, Saunderstown, Rhode Island

PICKLED BEANS FISHER

Pick (string if necessary) and wash 2 bushels of good Corn Hill (or similar variety) beans. Put in large kettle with water and cook until a bean taken between the thumb and forefinger mashes easily. Do not overcook. Drain water off and wash through enough cold water to cool beans and until water is clear and cold. Put ½ gallon very cold water in bottom of 8-gallon crock; add ½ cup salt. Put in 2 gallons of beans, press down firmly. Add another ½ cup salt and 2 more gallons of beans and pack well. Continue to add salt and beans, and pack, until all are used up. Cover with folded cheesecloth and tuck edges down around the sides. Place a fitted weight or heavy plate on beans. Cover jar with cloth and let stand for 7 days. After this time remove beans and pack in clean sterilized jars and cap tightly. Process in canner, remove, and store. *Elwood Fisher, Harrisonburg, Virginia*

END-OF-THE-GARDEN RELISH

1 pint fresh green or yellow beans
2 large carrots, cut into ¼-inch lengths
1 cup celery
4 white onions
1 cup each of red, green, and yellow peppers
1 small head cabbage
1 head cauliflower
4 green tomatoes
1 pint fresh lima beans
(2 cups corn)

Clean vegetables and chop fine, except for lima beans and corn. Cook vegetables separately or in combination until just tender. Drain and put into kettle. Pour the following mixture over them:

6 cups cider vinegar
4 cups sugar (part brown, part white)
⅓ cup salt
2 tablespoons mustard seed
2 tablespoons celery seed

Bring mixture to a boil, put in hot sterilized jars, and seal.

KIDNEY BEAN RELISH

1 small onion
3 stalks celery
2 hard-boiled eggs
2 cups dry kidney beans, cooked and drained
1 tablespoon mayonnaise
2 teaspoons relish (dill relish if you have it)
1 teaspoon curry powder
½ teaspoon salt

Chop onion, celery, and eggs and mix together. Add beans and mix in mayonnaise, relish, and seasonings. Chill before serving.

INDEX

ABOUT THE AUTHOR

John Withee spent his boyhood in Maine, where he worked summers as a market gardener, selling vegetables in Portland. He has been a bank messenger, a night janitor, a weather observer, and a door-to-door bakery salesman. He began working as a medical photographer at the Dartmouth Medical School, and spent 16 years as a photographer for Peter Bent Brigham Hospital in Boston. In 1976 he received the Professional Photographers of America National Award. Among other hobbies Mr. Withee has ridden freight trains across the country, flown twice to the Arctic in a Piper Cub, climbed Mt. Katahdin three times in winter, and explored the Rocky and Grand Teton mountains. No matter what else he has been doing at the time, he has managed to have a garden every summer.